Designing and Managing Complex Systems

Designing and Managing Complex Systems

David Moriarty

ELSEVIER

ACADEMIC PRESS
An imprint of Elsevier

Academic Press is an imprint of Elsevier
125 London Wall, London EC2Y 5AS, United Kingdom
525 B Street, Suite 1650, San Diego, CA 92101, United States
50 Hampshire Street, 5th Floor, Cambridge, MA 02139, United States
The Boulevard, Langford Lane, Kidlington, Oxford OX5 1GB, United Kingdom

ISBN: 978-0-323-91609-7

For information on all Academic Press publications visit our website at https://www.elsevier.com/books-and-journals

Publisher: Nikki P. Levy
Acquisitions Editor: Joslyn T. Chaiprasert-Paguio
Editorial Project Manager: Sara Pianavilla
Production Project Manager: Swapna Srinivasan
Cover Designer: Vicky Pearson

Typeset by TNQ Technologies

This book is dedicated to my father, David Patrick Moriarty, who, during my early years, sparked my interest in many of the subjects featured in it.

Contents

2.5 Sociotechnical systems

2.6 Consilient dynamics across scales

Part III
Creating and managing systems

3.1 Introduction to part 3

3.2 Structure and function

3.3 Capability and adaptive capacity

3.4 Engineering resilience

3.5 Assessing the system properties of your organization

Part IV
Case studies

4.1 Challenger and Columbia

4.2 Walmart, FEMA, and Hurricane Katrina

4.3 Lake Peigneur

Part V
Conclusion

Author biography

David Moriarty has worked as a medical doctor and as an airline pilot. Additionally, he has degrees in neuroscience and human factors and specializes in system design and complexity management. He works as a consultant specializing in these areas. Comments and feedback about this book can be sent to systems.thinking@hotmail.com.

Also by David Moriarty
Practical Human Factors for Pilots

Preface

Ashamed lest I appear to myself as a pure theorist, unwilling to touch any practical task.

Plato, Ep. 7, 328c

Planning a book such as this requires some effort. As well as deciding what you want to say, you also need to map out how you intend to say it. The first step is to write an outline. The outline includes a detailed description of the planned chapters with a summary of their content. By the time I was happy with the outline, I was on the 14th arrangement of my chapter list. I had spent a great deal of time chopping and changing the order to try and ensure that the concepts that I was planning to present would follow a logical order. Learning is, by necessity, a deliberate and sequential process, and when trying to explain different strands that will come together later in the book, you have to hope that the reader has enough faith that it will all make sense in the end even if it is not immediately clear as they work through the text.

To make matters worse, a book about complexity is bound to contain some complex ideas. The challenge for the writer is to present these ideas, not only in some sort of logical order, but also with sufficient detail so that the reader can mentally build their conceptual "house of cards" as they progress through the book. It was the search for this logical sequence, and the planning of how much detail to include that led me to try out 13 different plans before settling on the one that forms the basis of this book.

Suffice it to say, the field of complexity is a huge, ungainly entity encompassing knowledge from disciplines across the entire spectrum of human knowledge. As the saying goes, there is only one way to eat an elephant—one bite at a time. However, I would add that there is still a decent chance that it will be a messy process.

I do not claim to present a unified view of complex systems or a complete perspective on the subject. I am convinced that both goals are impossible as things stand now. At the end of this book, there will be no breathing of sighs of relief because you now have all the tools that you need to manage complex systems. However, rather than putting in as much as I can about the subject, I have tried to take a more focused approach.

Even before complexity science was recognized as a discipline in its own right, researchers and practitioners in other fields would either develop or uncover concepts that might ultimately add to the knowledge base of complexity. Sometimes ideas would flare up and capture the collective imaginations of a diverse group of academics before burning out and becoming all but forgotten. Other ideas would fizzle out almost immediately. Whether popular or not, all these findings are like rocks strewn across the bed of a shallow river. Some of these rocks will allow us to make progress crossing the river by setting us up to stretch our foot out for the next rock beyond. Some rocks will be loose and leave us soaking wet. In progressing to the other side, we must select a path whereby each rock we choose sets us up for the next one. There may be other stable rocks to the left and the right of our path, and there are other potential routes apart from the one we have chosen. However, we cannot step on every rock in the river during our crossing.

That is to say that in selecting the concepts that are included in this book, I have had to choose a specific combination that I think are complementary. It would be impossible to outline every topic that falls under the umbrella of complexity science as this would leave the book just a jumbled mess of ideas.

I should also explain why I chose the particular quote that I did to preface this introduction. If I had to pick one adjective that a reader might use to describe this book after reading it, I would hope that word would be "useful." The science of complexity is a fascinating one, and it would be very easy to engage and dazzle you with a plethora of stories concerning slime molds, merging galaxies, and the intricacies of the human brain. While I hope that the substance of this book will be interesting to you, I am far keener that the substance of this book will be *useful* to you. In trying to describe a framework through which we can understand, design, and manage complex systems, I have tried to select areas of study, which will come together to form as rational a description as possible. The domain of complexity science does cross many traditional academic disciplines, and it would be impossible to present a complete picture of everything in the space of one book. Instead, I have tried to curate the findings from various fields of study and crystalize those findings into something that is accessible and useful. Fellow researchers may well vilify me for not including material from one field or another (including the mycologists and their slime molds, although they do get a brief mention later), but this has not been done haphazardly. Given a choice between an unstructured book filled with fascinating stories and a structured book that presents selected concepts in a manner that takes the reader through a logical sequence, I choose the latter.

And so, to the matter at hand. This book will require some patience and engagement from the reader. It is divided into four parts, with each part building on the parts that came before.

- The first part looks at the history of one particularly fruitful approach to complexity and gives the reader an overview of the background and scope of complexity science.

- The second part looks at our current understanding of complex systems in various domains (physical, biological, technical, social, and organizational). It introduces the idea that there are similarities between the dynamics of successful systems across domains.
- The third part uses the preceding material to develop principles that can be applied to design and manage complex systems successfully.
- The final part focuses on case studies concerning failures and successes within complex systems. It demonstrates how the presence or absence of advantageous design and control principles can lead to success or failure.

While I could try and summarize my conclusions here, the value to the reader will be in being able to assemble the diverse parts of this particular academic jigsaw to reach the same conclusions themselves by the time they finish Part 4 of the book. From my teaching experience, I know that being told something is far less useful than working something out for oneself.

With the idea of utility in mind, this book is also designed to be accessible. As the book proceeds, you will be faced with an increasing amount of nomenclature, and so a glossary is included to allow you to find a reminder of the meaning of a particular term without having to go back through the preceding chapters. Occasionally, to recap on a particularly important set of concepts, a summary may be presented.

Before we begin, it is worth saying one more thing. Human beings aren't built for complexity. In Part 2, I will go into more detail about the limitations of the human brain, despite all of its amazing capabilities. We have created, probed, and discovered systems that change and move in ways more rapid and complex for us to keep up with. Even before the modern era, humans evolved an inbuilt imperative to simplify the deluge of stimuli that bombards us as we interact with the world. As the world accelerates, our simplification imperative grows as well. For the most part, this instinct to simplify works for us and allows us to function reasonably well as participants in this complex universe. However, we mustn't fool ourselves into thinking that our simplified understanding of a particular aspect of this world entirely matches reality. The model train set will allow the carriages to move between two points but won't replicate ticket pricing, driver's unions, safety programs, or government regulation. There will be times when we need to delve down into the granular structure of the systems that surround us in the full knowledge that a more straightforward representation exists on the surface. Our simplification imperative can make us want to shy away from the complex because to engage with it will require more cognitive capacity than we may be willing to surrender to the task. However, from time to time, this is the instinct that we must fight as we dive below the surface, for it is beneath the surface that the hidden dynamics unfold that determine what happens up above.

As I have stated, the science of complexity encompasses findings from a diverse number of fields. There is also a subtle but persuasive theory that the dynamics observed in one sphere of human inquiry may well be the same as those observed in another field, but that the increasing specialization of academic disciplines means that we are less likely to spot the crossover. My interest in this idea came initially from a book by the noted evolutionary biologist Edward O. Wilson. Published in 1998, "Consilience: The Unity of Knowledge" was an exploration of this idea with particular emphasis on the crossover that occurs between the sciences and the humanities. While the following book focuses more on the crossover between the natural sciences and the social sciences, especially management, Wilson's work has acted as the catalyst for my exploration of this subject. His book is a fascinating and illuminating read[1].

Consilience is probably a novel word for most people, but its structure may seem vaguely familiar, perhaps through its similarities to conscience, science, or silence. Indeed, even as I type, my computer is telling me that it is alien and unknown, being underlined in red, a sight familiar to inept typists such as me. However, the seed of this book was planted in my brain when I was first introduced to this concept after reading Edward O. Wilson's book.

I read Consilience shortly after it was published in 1998. Wilson's beguiling and compelling thesis was about the unification of knowledge across fields and suggested that there were universal dynamics that revealed themselves through the natural sciences, the social sciences, and through the humanities and that these dynamics may actually be the same across these diverse disciplines. Like different religions trying to uncover the same truths about God, the various spheres of human inquiry were revealing the fundamental dynamics of reality but were using different language to describe them. Consilience, or, in Edward Wilson's terms, the unity of knowledge, suggests that profound insight in one field of human understanding is likely to be replicated in another area and that natural sciences, social sciences, and the humanities share more similarities than most academics would care to admit.

Seeing these similarities becomes progressively harder and harder as our knowledge of these disciplines deepens and diverges. The growth of human knowledge necessarily means that academics have more to learn about increasingly specialized areas of study. Before the Renaissance, educated people could speak with equal fluency about the natural sciences as well as about literature, art, and music. A classicist may also be a scientist. Because of the dearth of source material from which to learn, any knowledge was precious, and an educated person made less distinction between the academic disciplines.

However, at some point, our access to stored knowledge became easier, and the depth of that accumulated knowledge reached a certain critical mass whereby we had to give up on the idea that we could learn it all. Limitations of cognition meant that we either had to know a little bit about a lot or a lot about a little bit. It could be argued that this was the inflection point after which we began to lose the opportunity to see how those areas of study related to each other. For example, careful microscopic examination of the wood does not allow us to see how the tree fits into the forest ecosystem.

In his 1959 book "The Two Cultures," CP Snow argued that the divergence between the arts and the sciences was stymying the application of knowledge to solve global problems[2]. In later works, he also included the social sciences as a third culture that had also diverged from the first two. Snow argued that it was not just that academic inquiry had divided into separate fields out of necessity. Instead, those following a particular academic path (or a particular "culture") were also likely to view other cultures with suspicion. Those who study the humanities might look with disdain at the perceived illiteracy of the natural scientists. Those in the natural sciences might be appalled at the scientific ignorance of those in the humanities. Rather than accepting that we necessarily have to give up certain avenues of study due to limitations of time and cognitive capacity, the emergence of these separated cultures has also created different academic factions. Their entrenched suspicion of those outside of their particular group not only limits transdisciplinary understanding but creates different academic languages, which make it difficult to understand each other's worlds even when the opportunity to collaborate does arise. It is a viewpoint that led directly to Edward O. Wilson's book.

At this point, I must hold my hands up and say that I, too, am a product of this gradual academic pigeon-holing. From age 16 onward, I had to nail my colors to the flagpole and accept that I would go down the scientific route. Even though I enjoyed other areas of study, limitations of time and my planned career for the future meant that there was only so much I could study, and so subjects that may well have been a help to me, in the end, had to fall by the wayside. Although my early focus was on the biological and medical sciences, I tried to maintain an enthusiastic but admittedly amateurish interest in other areas of study, which will hopefully prove sufficient for this book's purposes. It will also become clear to the reader that some of the material on which this book is based is derived from safety-critical industries. The reason for this is simple. It was academics in the field of industrial safety that made some of the important breakthroughs in how we view complex systems as they attempted to find ways to understand and manage them more successfully. Some of the examples used in the book come from safety-critical industries such as aviation or nuclear, but whether your organization is safety-critical or not, the principles are the same. It is not just safety failures that stop a complex system from working, and the dynamics that cause commercial failures often have similar roots in the underlying mechanisms of complex systems.

More importantly, I think we have been hamstrung by the increasing specialization of academic inquiry and our intransigence to look outside our narrow area of study to find fruitful help in other disciplines. Throughout history, human endeavor has led us toward unpicking the complex. Neurobiology, quantum physics, cosmology, macroeconomics—no complex system is immune from our relentless inquiries. But when we consider the broad range of systems that exhibit complexity, despite the plethora of disciplines through which these systems are typically investigated, they all have one characteristic in common—they are not easy to understand, i.e., they are complex. We seem to constantly have our noses pressed up against the translucent windows of engine rooms containing these complex machines. We can see the outline of some of the cogs and gears, and we can hear the pitch and volume of the motors driving these dynamic systems. Some people have managed to see through a keyhole or through a chink in a door to get a fleeting glimpse of the engine that drives the machinery of complexity, but the grand blueprint remains hidden. Is there a component list and a set of instructions for these machines? Can the components and their interactions be found driving complexity in different types of systems? And, crucially, what can we learn about this general type of machine that could help us manage the machines driving the complex systems in other fields? The complex dynamics generated by these principles affect biological, physical, social, and environmental systems and even phenomena that occur in the worlds of the humanities, such as visual art, music, and literature. A physicist may be beginning to discern the particular workings of the complex system that she is studying without realizing that these insights are precisely what are needed to give the sociologist insight into the problem that she is attempting to solve.

The concept of consilience will be one of the pillars on which this book stands. Perhaps you are a specialist in one particular academic field. Maybe you are someone who is tasked with managing or participating in some sort of complex system. To make this book as useful as possible, I will try and cater more to the latter but with enough meat to satisfy the former. I hope that this balancing act will help me construct something that is engaging enough for the general reader. As we look at different areas of the book, I will occasionally highlight a concept or phenomenon in a box as I have done below:

> With consilience in mind, I think some specific observations from a particular field might be generalizable. I might not necessarily be able to generalize it myself across lots of different fields, but an astute reader may find that the highlighted concept gives them an insight into how things might work in some other discipline. These boxes will mainly be a feature of Part 2.

Finally, to give you as much of an understanding as possible about how everything fits together, here are the other pillars aside from consilience that this book is going to stand on:

- The Cynefin framework
- Self-organizing systems
- The Viable System Model
- Resilience Engineering

There is no reason to jump ahead or just to focus on these topics in isolation. The goal is to show how these subjects fit together in a consilient and, hopefully, useful manner. From the outset, I want to clarify that these are the really big, stable rocks that I have identified on our path across the river. But remember, the little rocks are important too.

Complexity isn't a vice.

CJ Cregg – A character from the TV series, "The West Wing"

Acknowledgments

I'm grateful to David Beausang, Jimena Gonzalez Jimenez, Ben Rose, Christopher Attwood, and Michelle Wood for their help and advice as I was preparing this book. I'd also like to particularly thank Dr. Robert Moir for his incredibly useful insights into the world of physics. I am also grateful for the help and support of all the teams at Elsevier especially Joslyn Chaiprasert-Paguio and Sara Pianavilla.

Part I

Cybernetics

Chapter 1.1

Control and communication

I think the next century will be the century of complexity.

Stephen Hawking, physicist

The room was hexagonal, and its walls were paneled with darkly grained wood. Interspersed around the walls were displays of varying designs that could be controlled from any of the seven swivel chairs arranged in the center of the room. These chairs would not have looked out of place on the bridge of the USS Enterprise. Kirk's Enterprise, though, not Picard's. Low-backed and with the seat slightly tilted upward, they owed some of their design aesthetic to Charles Eames. However, the right armrest was widened and had been extended to house a control panel that would operate the screens.[1] The room had been designed with a grand purpose in mind; it was to be the command-and-control center for the economic machinery of an entire country. It was the apotheosis of a bold field of scientific inquiry that had begun in the 1940s, a discipline that had been created and nurtured by some of the finest minds of the 20th century. It was also the grand hope of the new socialist regime in Chile and would potentially be the bedrock for a tremendous economic resurgence throughout the country. Then, on September 11, 1973, the socialist government of Chile was violently overthrown by General Augusto Pinochet. In La Moneda, the presidential palace, President Salvador Allende shot himself rather than be captured. Shortly after that, Pinochet's troops brok into the control room and destroyed it all.

In 1906, Norbert Wiener graduated from high school in Massachusetts and entered the prestigious Tufts University in downtown Boston. He was 11 years old. The newspaper, The World, hailed him as "the youngest college man in the history of the United States" and, with that, Wiener became the most famous child prodigy in America. His father was a Professor of Slavic Languages and had homeschooled his son according to a system of his own devising. Unlike many child prodigies who suddenly find that their youth is the only thing that separates their intellectual achievements from those of their older contemporaries at university, Wiener's academic progress was stellar. He went on to achieve his Ph.D. at 17, spending time studying under Bertrand Russell and G. H. Hardy at Cambridge, and the legendary mathematician David Hilbert at Gottingen. Although his primary focus was mathematics,

Designing and Managing Complex Systems. https://doi.org/10.1016/B978-0-323-91609-7.00001-9

Wiener also studied philosophy, biology, and engineering, subjects that would prove essential to much of his later work. In 1919, he became a lecturer at the Massachusetts Institute of Technology (MIT). He rose quickly to become a full Professor of Mathematics, a position he held for the next 40 years.[2]

Short, a little rotund, and extremely myopic, Wiener was almost the perfect archetype of the absent-minded professor. This image was completed by thick-lensed tortoiseshell glasses and a Van Dyck beard. A notoriously bad driver, his MIT colleague Paul Samuelson remembered that his wife had expressly forbidden him to accept a lift home from work from Wiener despite them living in the same part of town. A late husband, she surmised, was better than a *late* husband. One particular story recounted by one of his colleagues concerned the day that the Wiener family was moving to a new house. Before he set off for work, Wiener's wife Margaret reminded him that they were moving today and that he should remember to go to their new address at the end of the day. Assuring her that he wouldn't forget, Wiener promptly forgot. Returning to his old address that evening, he was surprised to find that the house was empty and so asked a little girl who was playing in the street if she knew where the Wieners had moved to. "Yes daddy," she replied. "Mom said I was to stay here so I can bring you to our new house."[3]

By the 1930s, Wiener had established himself as one of the academic stars in MIT's already glittering firmament. Although part of the mathematics department, he was a true polymath and could be reliably called on to contribute to cutting-edge research in many other scientific disciplines. He had a tradition of taking rambling "Wiener-walks" around the campus each day, stopping passing professors and students alike to expound on whatever idea or problem was concerning him at the time. On one occasion, he stopped a young physics student who was walking in the opposite direction. After their conversation, Wiener asked the young man what direction he had been walking when he had stopped him. "You were going toward Building 8," the student replied. "Thanks," Wiener said. "That means I have already had my lunch."[2]

The outbreak of World War II brought a fresh surge of interest in the practical applications of new scientific thinking. If World War I had been a war of attrition, the outcome of World War II would be decided according to which side could best harness cutting-edge science to gain a competitive advantage. The vice president of MIT, Vannevar Bush, had been selected by the US government to set up a new organization that would do just that—the National Defense Research Committee (NDRC). With full access to the country's top scientific minds, Bush was in a position to handpick leading scientists to work secretly on government defense projects. One such project was clearly going to require an immense effort and needed a first-rate mind to lead it. Fortunately, Bush knew the right man for the job, and Wiener was brought in to lead the project.

The challenge involved a new use for a new technology—radar. The military wanted to see if there was a way of using radar in conjunction with a statistical computation system to predict the future flight path of highly maneuverable warplanes. This information would then be instantly fed to the control unit of a bank of antiaircraft guns to shoot these aircraft from the sky automatically. The ranges over which these weapons were designed to work meant that you could not simply target an aircraft and shoot it down. By the time the shell reached the targeted area, the aircraft would have flown on a considerable distance. Instead, the guns had to fire shells to where the aircraft was *going* to be, taking into account the distance the shell would need to be fired and the expected flight path of the aircraft during that time. The challenge was considerable. It would have been difficult even if the pilots could be relied on it keep a constant course and speed. However, the pilots of that era were trained to undertake random evasive maneuvers if they were under fire. Wiener's task was to create a way of predicting how these evasive maneuvers would unfold during the course of an assault so that the antiaircraft weapons could be targeted not to the patch of sky where the aircraft was, but to the patch of sky where the aircraft would be. Much like the efforts to crack the German wartime codes underway at Bletchley Park in the United Kingdom, Wiener and his collaborators needed some way into the problem of how the complex decisions that the pilot must make under fire resulted in their choice of evasive action.

The answer came in the revelation that the options available to the pilot were more limited than they had first thought.[4] Traveling at high speed, the pilot can only make certain turns at specific rates. If he tries to turn faster than this or tries to turn in a direction that changes the aerodynamic load on the aircraft too much, he risks overstressing the aircraft or losing consciousness himself from the G-forces. By setting up a reciprocating information flow between the artillery system and the radar, the pilot's evasive actions following the first round of antiaircraft fire can be used to refine the second. The radar signals would update the statistical program to narrow down the possible areas where the target would be next. This process would continue until the target was shot down or it escaped.

Wiener had laid out the principles by which such a system could theoretically work. He now needed to embed those principles in a machine that would do the job. Firstly, the radar signals were fed into a makeshift machine that performed the calculations. Next, the results of these calculations would be fed directly to the gun turret that would then fire the shell. If the shell missed, the process would continue through another cycle, and the calculation would be able to factor in that the change in the flight path that had allowed the pilot to avoid the first shell left him with fewer potential flightpath changes available to him, which improved the chances that the second shell would hit the plane. Over subsequent cycles, the system was designed to progressively reduce the gap between where the shell exploded and where the aircraft was.

Sometimes, during the course of testing, Wiener and his colleagues noticed that the gun turret would exhibit some unusual behavior. Rather than progressively settling on the predicted target area, it would swing wildly to the limits of its allowed range of movement. A neuroscientist colleague of Wiener's, Arturo Rosenbluth, noted that this was similar to the tremor that a patient with Parkinson's disease experiences when they reach for an object. In these patients, it was common to see a slight but constant tremor in the hands when the patient was at rest. However, when the patients reached for an object or carried out a specific manual task, this tremor would become a lot more pronounced, almost to the extent that the arm and the wrist would move up and down to nearly their fullest extent. The neurological problem had been traced back to a fault in a set of circuits toward the back of the brain that would typically coordinate fine movement, i.e., the process of progressively reducing the difference in position between the desired object and the hand that is being extended to clasp it. Having realized that a mechanical circuit that carried out the same function would eliminate the gun turret's "tremor," Wiener was able to refine the system, and trials showed that the mechanism could predict the path of a highly evasive aircraft up to a second before it got there. However, it soon became apparent that there was a problem. The calculations themselves were not an issue. The issue was the time available to complete them. Computers were a nascent technology in the 1940s, and even bringing the most advanced systems to bear on the problem couldn't help solve the equations quickly enough. Wiener had proved in principle that this sort of system was possible, but without the computing technology needed to put it into practice, the project was ultimately scrapped.

The details of this secret undertaking were only declassified some years later, but the work Wiener carried out during that period turned out to be some of his most profound. In a rather technical report that he submitted to the NDRC entitled "The Extrapolation, Interpolation and Smoothing of Stationary Time Series,"[5] he unwittingly laid the foundation for the new science that would be his greatest legacy. This paper defined the concept of control, particularly control in complex systems, and elucidated a phenomenon that had been underappreciated up to that point: feedback. The reciprocating nature of the communication between the radar sensor and the predictor program was essentially based on a feedback loop that progressively reduced the error between the plane's predicted position and the plane's actual position. Using information from the world of neurophysiology, Wiener had also defined the means of purposeful control through negative feedback. In the case of the NDRC project, this was the smooth movement of the gun turret to precisely the correct position as dictated by the predictor program. It is clear that these ideas of feedback in control systems persisted in Wiener long after the project had been canceled and, particularly, how the world of biology had elucidated valuable concepts in the world of engineering. These worlds would come together in a far more profound way a few years late.

In a New York hotel, in March 1946, the first Josiah Macy Jr. Foundation conference to happen since the end of the war started. In the 4 years previous to this, Wiener had worked on his ideas regarding control in biological and technological systems in collaboration with a range of scientists, all at the very peaks of their fields. Among the attendees were Warren McCulloch, a bohemian, brilliant neurophysiologist from Chicago; Gregory Bateson and Margaret Mead, the famous husband-and-wife anthropologists; Arturo Rosenbluth, the neuroscientist who had helped solve the problem of the erratic gun turret; and John Von Neumann, a prodigy, like Wiener, whose expertise spanned computing, engineering, mathematics, nuclear physics, and obscene jokes. All told, the conference included internationally recognized experts from the following fields: mathematics, engineering, sociology, psychology, neurophysiology, physiology, philosophy, ecology, anthropology, neuroanatomy, and medicine. This would be the first time that they had all gathered for the express purpose of discussing this new discipline. This Macy Conference was the first of ten that would focus exclusively on this new science and on the increasingly apparent commonality that academics from radically different fields were using to describe the phenomena they were investigating.[6] Wiener was always central to these meetings, and, as was his right as the prime mover, his words were accorded particular significance.

The Macy Conferences that took place between 1946 and 1953 were arguably the most significant transdisciplinary conferences that ever took place judging by the effect they had on the world as we know it, but also because of the unparalleled brilliance of the participants. In the early meetings, as the participants were all trying to find their way into this mysterious new field that they were creating simultaneously as they were exploring, the clashes between them were spectacular. Warren McCulloch, the only man brave enough to chair these meetings, talked afterward about the brutal language, the tears, and, sometimes, the threats of violence as these prestigious scientists attempted to flesh out their positions while being interrogated to the point of breakdown by everyone else. It was McCulloch who recorded a quote attributed to Wiener used at one of these meetings: "Don't stop me when I am interrupting!"[2]

The first meeting laid the foundation for the meetings that were to follow. It started with the revelation shared by Wiener and Rosenbluth; that there appeared to be a set of principles of control and communication that were the same in machines and humans. To understand the significance of this insight, you have to know that the concept of negative feedback did not exist in the 1940s. We are all accustomed to the idea almost instinctively because we now see how it pervades nearly every system designed to control some parameter or other. It could be a thermostat controlling the temperature in a room or the human endocrine system controlling the level of sugar in the blood. Both of these systems work through negative feedback, but, at the time, not only was the concept of feedback not well understood, there wasn't a framework through which to explore it. Control and communication in systems, whether mechanical or biological, were not well defined. The early Macy Conferences elucidated these concepts and laid the foundation for an entirely new scientific discipline that could systematically explore them.

As the undisputed leader of this field, Wiener was to write the definitive book on the subject. All the subject needed was a name. As the book's submission deadline was fast approaching, Wiener eventually settled on a name that suited him. Given that the subject was devoted to the science of control, he appropriated the Greek word for steersman: kubernetes. After a little bit of anglicization, the new science had its new name: Cybernetics.

Wiener's book, "Cybernetics: or Control and Communication in the Animal and the Machine," was published in 1948 and was dedicated to Arturo Rosenbluth, the neuroscientist who described those same phenomena at work in the animal world that Wiener realized could apply to the world of machines.[4] The sweeping range of profound ideas that it contained made it one of the most important scientific books ever to be published, and, indeed, it did change the world. Homeostatic mechanisms (those physiological processes based on negative feedback) were not just found in the realms of biology but could also be found and, more importantly, put to work in technological and social systems. The book laid the foundation for information theory and for the concept that messages could be transmitted logarithmically in binary units, soon to be called bits. He postulated that certain psychiatric disorders could be traced back to problems with normal feedback systems within the brain and that the new science of cybernetics could be brought to bear on computing devices and large-scale social networks. The book was a sensation and went through five printings in 6 months. It also made Wiener one of the most famous scientists in the world. Cybernetics had arrived.

When we look at the popularity of cybernetics at that point in history, it may be somewhat difficult to understand why it had such an effect on people. We live in a society where the mechanisms of communication and control are so engrained into our daily lives that it may be hard to imagine a time when people didn't have such an intuitive understanding of the principles that underpin them. Most people who have heard the unpleasant feedback that occurs when a microphone is held too close to its speaker will know that it is the result of a noise being picked up by the microphone, amplified through the speaker, picked up by the microphone, further amplified by the speaker until something interrupts the positive feedback circuit. The principles of cybernetics and information theory are encoded into the microchips and circuits that form the basis of 21st-century society, and it is only in a world where computers are ubiquitous that we can have such control over our environment. To understand how revolutionary the ideas of cybernetics were, you must imagine the world of the early part of the 20th century; a world without computers and with rudimentary and fragmentary channels of communication; a world of simple electronics accessible only to small numbers of people. It was understood that there were complex entities in that world, most prominently the human body and, in particular, the brain. There was a vague understanding that the complex functions of entities such as the brain resulted from sets of interdependent components working together in some sort of harmony. But it was the nature of this interdependence and the precise way that components communicated with each other that had eluded formal definition up to this point. Cybernetics wasn't so much a science, but more so a language that allowed scientists from various disciplines such as engineering, biology, sociology, to explore and delineate the dynamics of the complex systems they studied.

Across the Atlantic, this new thinking sparked an explosion of debate and investigation. Biologists, physicists, and economists now had a common language to describe the behavior of the systems they studied. Of the European groups that formed after the publication of Wiener's Cybernetics, it was a small group, named The Ratio Club, started by John Bates, a neurologist at the National Hospital for Neurology and Neurosurgery in London, which most determinedly took up the challenge of making headway into this new field.[7] In his letter to prospective members, Bates alluded to the incipient sense within the scientific community that a new, unifying perspective on reality was on the horizon even before cybernetics was formally introduced. As he put it, the club was "for those who had Wiener's ideas before Wiener's book appeared." Unlike the Macy Conferences, the other criteria for admission to the club were that you were young, smart, and not a professor. Bates wanted an informal, open atmosphere and felt that senior academics carrying the weight of the previous accomplishments would stifle this. Following his letters of invitation, the Ratio Club swelled by another 20 members, most of whom validated Bates' recruitment profile by going on to have outstandingly productive careers. One member of the group was the 37-year-old Alan Turing who, by the club's inception in 1949, had already helped the allies win World War II by designing the machine that cracked the German Enigma codes, an accomplishment that was kept secret until the 1970s. It was at one of the meetings of the Ratio Club that Turing expounded on his idea of machine intelligence that might one day pass the test of fooling a human into thinking that they are interacting with another human, the so-called Turing Test, which he expounded in a scientific paper that included a section called "The Imitation Game."[8]

Another founding member of the Ratio Club was a psychiatrist named Ross Ashby. It was Ashby who became the most significant cybernetician in the group and contributed to this new and emerging field by authoring two books: "An Introduction to Cybernetics"[9] and "Design for a Brain."[10] However, Ashby's contribution to the field was not merely theoretical. In 1948, even before the founding of the Ratio Club, he had constructed a machine that Time magazine said was the closest thing to a synthetic brain so far designed by man.[11] The mechanism was called a homeostat and was designed to demonstrate the principle of ultrastability, namely the ability of a system (in this case, a mechanical one) to adjust to external disturbances and recover its original state. The machine was built with army surplus parts and was electrical in nature. Four modules were interconnected in such a fashion that disturbing the electric output in one module, for example, by moving its conductive unit, would cause the conductive unit in another unit to move in the opposite way to compensate. This, in turn, affected the third module's conductive unit, which then affected the fourth. The initial disturbance would ripple throughout the rest of the system. However, the connectivity between modules was such that negative feedback between them dampened the effects of these disturbances. Progressively, as information in the form of electrical signals was propagated throughout the system, the negative feedback eventually returned the system to its original neutral state. Thus, Ashby had demonstrated that negative feedback in a mechanical system could enable it to maintain a steady-state despite external disturbances. The same "homeostatic" mechanisms have been shown to underpin biological systems, allowing them to maintain body temperature, blood sugar levels, hormone levels, etc. In 1952, Ashby was invited to New York to demonstrate the homeostat at the ninth Macy Conference.

Ashby's other significant contribution to cybernetics was his description of a fundamental principle of successful control. Ashby realized (and proved mathematically) that a system's controller needs to have at least as many states as the system being controlled. To take the example of his homeostat, Ashby suggested that the reason that the mechanism could achieve stability was that the control unit could issue at least as many different types of command as there were types of disturbance. To use a more contemporary example, a simple on/off switch would not be appropriate to control the output of a speaker connected to a microphone. Suppose you imagine one of those vast mixing panels used in music studios and in concert venues to control the output coming from the musician's microphones. In that case, there are many states that the panel can be set to give the best output from the microphones. Ashby used the term *variety* to describe the number of states a system can take, and his law that the controller of such as system should have at least as many states is known as *the law of requisite variety.* This important concept will be covered in more detail later.

Ross Ashby was known to be somewhat dry and cerebral but was undoubtedly the leading light in cybernetics in Europe in the mid part of the 20th century. His psychiatric training and interest in understanding the dynamics of the brain led him into this field dedicated to control and communication in mechanical and biological systems. His expertise in cybernetics eventually led him away from clinical medicine, and he ended up as a professor of electrical engineering and biophysics at the University of Illinois. The story of how he got this job gives some measure of the man.

In 1960, Ashby was at a conference in Illinois with the leading lights of cybernetics at the time. One of the professors at the University of Illinois, Heinz von Foerster, was trying to convince another cybernetician, Stafford Beer, to take up a professorship there. Beer was an Englishman with a young family back in England and was not keen to move them to the United States. Von Foerster then tried to get Ashby to convince Beer to take up the professorship. Ashby demurred, stating that he wasn't going to convince Beer to do anything, particularly as he himself had never been offered a professorship. When Beer recounted the story, he said that a hush fell upon the collected group. It had been assumed that because of Ashby's preeminence in the field, particularly among the European proponents of cybernetics, he had had offers by the truckload but had turned them all down.[3] When Von Foerster realized that this wasn't the case, he formally offered Ashby the professorship. Immediately, Ashby accepted and asked to use the phone to call his wife back in the United Kingdom. The story goes that when she answered, he said, "we're emigrating. Sell the house."

Later that same evening, as Beer and Ashby walked to the university's faculty club, Beer said that he was surprised that a man like Ashby would make such an important decision so quickly without weighing up all the pros and cons. Ashby was aggrieved that a fellow cybernetician would take such a view. He stated that investigating and mapping out all the pros and cons of such a decision would be an immense, time-consuming task that might never be adequately resolved. In such a situation, he said, there were only two options. First, when Von Foerster made him the offer, his brain immediately said yes. Ashby knew then that subtle mechanisms at work in the brain could give solutions to problems even without conscious consideration. In this instance, knowing that a rational analysis would be time-consuming and incomplete, he went with the immediate answer that his brain came up with. When Beer asked him what the second option would have been if his brain hadn't immediately provided him with a solution, Ashby replied, "toss a coin, of course!" When Beer thought about this sequential process of decision-making options, he had to respect its intense rationality.

When it came to practically applying the lessons of cybernetics, it is probably Stafford Beer who had the most success and, indeed, the most ambition. Slightly younger than Wiener, McCulloch, and Ashby, his path into the world of cybernetics had been a circuitous one. Born in London in 1926, he dropped out of university to join the army. Eventually, he was assigned to the branch of the military concerned with operations research, where he became interested in processes and organization.[12] It was inevitable that this interest would lead him to the research being carried out in the new field of cybernetics, and Beer would eventually become one of its leading figures. After leaving the army, he rose to a senior position, Director of Operational Research and Cybernetics, in the largest steel company in Europe. From there, he launched into a successful and lucrative career as a management consultant, applying the principles of cybernetics to large, well-paying organizations. He published several books and many papers on the subject and, during the late 1960s, began to derive a model that he thought would encapsulate the cybernetic principles required for a successful, adaptable system. This was to be his principal and most lasting contribution to the field of cybernetics—the Viable System Model (VSM).

The VSM was a breathtakingly intuitive leap. Beer realized that the most successful system controller in existence was the human brain supported by the body's nervous system. Using this as his model, he mapped out the lines of control and communication that were required to allow the brain, and thus the person, to survive and thrive in a highly dynamic world. This was one of the great attempts to understand neurological complexity using the language of cybernetics, and the VSM was the result. It is worth looking at this model in detail as it embodies many of the principles required for managing complexity. The VSM defines five major groups[13]:

- System 5 – Responsible for system identity and high-level policy.
- System 4 – Uses environmental data to predict how the system must adapt to remain viable.
- System 3 – Can be defined as operations planning and control. Establishes rules and distributes resources to System 1. (System 3* represents the audit functions of System 3, i.e., how information regarding what is actually being done at System 1 level is communicated to Systems 4 and 5).
- System 2 – The information channels that entities at System 1 level can use to organize themselves, e.g., allocating shared resources, communicating needs and requirements between departments, etc.
- System 1 – A large organization will have many System 1 entities representing individual functions. In an airline, departments such as fleet management, engineering, training, flight operations would all be System 1 entities. It is important to note that System 1 entities are in themselves viable systems and so will have their own Systems 2–5. In this way, the VSM is often called a recursive model as System 1 entities may have a similar structure to that of the organization as a whole in terms of policymaking, adaptive capacity, planning/control, and information channels.

Systems 1 and 4 have direct interaction with the environment (the physical, commercial, and sociological contexts within which the entire system is embedded), and the whole VSM can be shown diagrammatically along with the nervous system structures on which it was based (Figs. 1.1.1 and 1.1.2):

Toward the end of the 1960s, Beer became interested in applying cybernetic principles to sociological and political systems. These principles had already been extended across mechanical and biological systems, and Beer theorized that they would hold for more hierarchical, less predictable systems and even for society as a whole. His work in this area resulted in his being approached by the new socialist government in Chile. His task would be to design a cybernetic system that incorporated his Viable System Model to manage the entire country's economy. It was intended to herald a new era of socialist, technology-driven governance and would be called Project Cybersyn.

To receive information from industries distributed along such a long, narrow country as Chile, each factory would have a communications terminal installed. These would then communicate each factory's input and output data

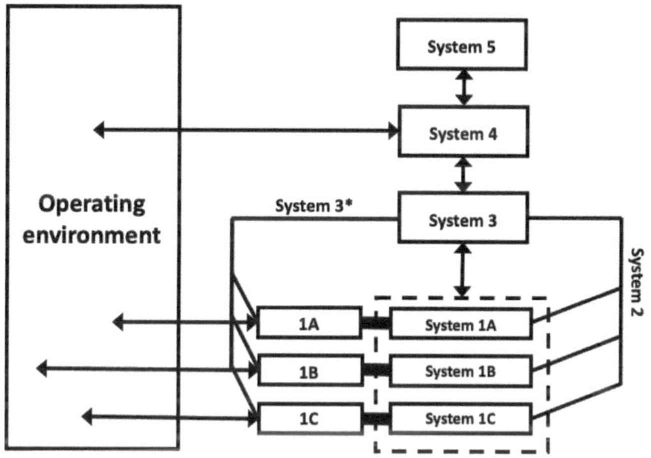

FIGURE 1.1.1 The Viable System Model.

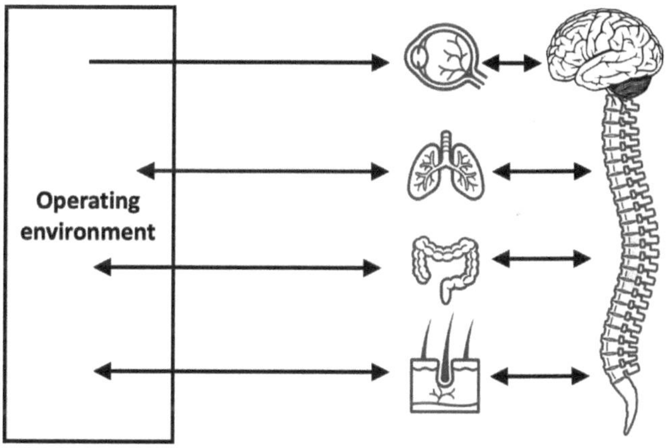

FIGURE 1.1.2 Corresponding arrangement of the human nervous system. *Images copyright takito and graphixmania (www.shutterstock.com).*

to the central control room. These data would allow controllers to identify where there were shortages or surpluses of materials/resources and allow for the prompt and efficient remediation of problems and shortages. A new computer program was designed to manage this inflow of data and integrate it into a visual representation of the Chilean economy to be presented in a manner that was simple to understand. If there were a problem at one level of the industrial system, it would be communicated to the next level up for resolution. If that level failed to resolve the issue, the signal would be transmitted still higher. In this way, a pathway of accountability ensured that important signals highlighting problems would not get lost in the data flow.[13]

The physical communication network, the computer programs, and the actual control room all took time to construct. As this was the first incarnation of a practical economic control system based on cybernetics, there were many problems to overcome and questions to answer. Reassuringly, the incomplete system proved its worth early on when the as yet unfinished communication network was able to coordinate information to and from the central government that allowed it to mitigate the effects of what could have been a crippling trucking strike throughout the country. The information transmitted back to the capital highlighted where help was needed most to avoid paralyzing the entire economy.

Work continued at a frantic pace throughout the early 1970s. President Salvador Allende, the first democratically elected socialist head of state, was very enthusiastic about the project and hoped it would allow the government to gain some measure of control over the ailing economy. However, the project's development was beginning to be outpaced by the country's economic problems, the high rate of inflation, and growing disquiet in the population. Eventually, in 1973, General Augusto Pinochet led a military coup and wrested government control away from Allende. President Allende committed suicide on September 11, 1973. Stafford Beer was not in Chile when the Cybersyn control room was destroyed.

It is hard to describe what happened to the field of cybernetics after this. Some of the early researchers in the field started considering second-order cybernetics, the recursive application of cybernetics to itself, and reconsidering the role of the observer in the process of analysis. However, second-order cybernetics never really had the same impact as the original work, although this book does reexamine the role of the observer in more detail.[12]

Cybernetics continued to fall out of mainstream academic consciousness throughout the late 1960s and early 1970s. By 1973, Norbert Wiener had been dead nearly 10 years; Ross Ashby had died the year previously, and Stafford Beer was still mainly working outside of an academic setting. For a scientific discipline to thrive, there needs to be an academic structure to support it. Because cybernetics was a science of the 20th century, many traditional academic institutions simply tried to integrate the discipline into existing departments rather than creating a department around it. As such, when the leading figures in the field began to retire and die, the field itself also seemed to diminish. Many of the lessons and discoveries were integrated into different scientific disciplines, but as a standalone subject in itself, cybernetics faded from academic consciousness. There persist groups around the world that are still interested in cybernetics for its own sake and as a lens through which we might better understand the complexity of the systems surrounding us. However, the cybernetic models that are still in use are somewhat constrained by the fact that they are based on research that is more than 50 years old. For example, the VSM was based on the understanding of the human nervous system in the 1960s. We know so much more about it now.

For myself, I think the problem is also one of specialization, something we looked at in the introduction to this book. Cybernetics owes its existence and value to a small group of highly intelligent people, people who could work across traditional disciplinary boundaries to find the similarities in disparate technical languages. Norbert Wiener, Ross Ashby, and Stafford Beer were prime examples of this kind of scientist: both specialist and polymath. The nature of contemporary society and modern educational methods means that scientists with such a broad base of expertise are far less likely to arise. While we may be well able to understand the value of specialization, taken from a different perspective, modern educational methods encourage us to know more and more about less and less. This may serve to expand the breadth of human knowledge, but it limits our efforts to synthesize knowledge across academic boundaries. Knowledge may grow, but the consilience of that knowledge becomes less likely. To tackle complexity, we need to be able to employ the language and techniques of many fields. The lessons learned in neurobiology can be the ones that solve a problem in electrical engineering; the insights of a physicist can be the solution to an issue in sociology. The remainder of this book looks for such lessons across a range of academic domains to equip us with the tools we need to manage complexity successfully.

Part II

Learning from systems

Chapter 2.1

The simplification imperative

For every complex problem there is an answer that is clear, simple, and wrong.
H. L. Mencken, journalist

Chapter outline

2.1.1 Model making

It's a messy world. People, systems, machines, animals don't always behave as we expect them to. Few processes are 100% accurate 100% of the time. We frequently must contend with scenarios that we haven't had prior experience of and are forced to improvise solutions by cobbling together some strategies that might have worked for similar scenarios in the past. Not only that, but our own cognitive abilities are limited. The electrochemical transmission within individual neurons is quite slow. The world, especially some of our most recent additions to it, such as computers and vehicles, can change at a far quicker rate than our meager cognitive abilities can keep up with. Fortunately, human beings have evolved techniques that allow us to mitigate some of the messiness we face daily. For the most part, this is sufficient to keep us alive and intact, but it doesn't always work to our advantage.

Let's consider a human sitting at a bus stop waiting for a bus. Surrounding her is a complex, dynamic reality of people and traffic superimposed on a background of the physical environment (buildings, weather, etc.). She interprets the physical reality through her five senses. It has been estimated that the eyes absorb about 120 megabytes of data per minute.[1] Processing all this information would likely exceed the cognitive capacity of the brain, and so this information stream is reduced by using the processes of selective attention and the formation of perceptual models.

Designing and Managing Complex Systems. https://doi.org/10.1016/B978-0-323-91609-7.00002-0

We don't usually pay equal attention to all aspects of our physical environment. As I type this, I am aware of the room around me, but my attention is focused on my computer screen. A sudden movement in my peripheral vision may make me shift my attention away from the screen. Still, under normal circumstances, I am only dimly aware of the things outside of the immediate focus of my attention. The processes of maintaining attention and of building a perceptual model of the environment are intertwined. As we interact with the world, we are not making our decisions based on the true nature of that reality but rather on our perceptual *model* of reality. If we only acted based on the true nature of the reality around us, we would never knock over wine glasses or miss a step as we climb the stairs. It may seem like a subtle distinction but these capabilities, firstly of being able to focus and shift attention and then using this narrowed down information stream to form a perceptual model of the world, allow us to function without being overloaded by the massive waves of stimuli that engulf us every day.

It may be disconcerting to know that your perceptual model of the world is not as rich or detailed as you might imagine. Humans tend to build perceptual models using a little bit of real-time information from the senses and a lot of filler dredged up from memory. You may well have seen images or videos of perceptual illusions which take advantage of this hardwired tendency to create a model of the world as we think it should be rather than how it actually is. Although such illusions become more pronounced (and, often, more critical) during dynamic changes of three-dimensional objects, even a simple two-dimensional, static drawing can trick us into seeing the world incorrectly. There are hundreds of these perceptual illusions all over the Internet, and they've been used to try and uncover some of the programming that underlies our perceptual systems.

> Creating models is an essential process that allows humans to use as little cognitive capacity as necessary to create a sufficiently detailed framework to enable successful interaction with the world even without it being totally accurate.

There are three ways that this simplification imperative manifests itself in humans, and we will look at each one in turn:

1. Chosen simplification
2. Forced simplification
3. Entrenched simplification

2.1.1.1 Chosen simplification

We see this same drive toward simplification at every level of human behavior. Language simplifies reality and allows us to refer to an object or describe an action without having to show it or perform it physically. Maps are a simplification of real-world geography because it is necessary to reduce the scale of representation and lose some of the resolution so as to make a map that is portable. Cities often portray their underground metro systems in a manner that emphasizes the sequential nature of stations on a particular line rather than their geographical proximity. Economic modeling allows governments and central banks to set policy even though these models don't fully reflect the complexity of the underlying financial markets.

Creating models necessarily means losing some detail. The rougher the model, the less detail it contains. Depending on what we are hoping to use the model for, there may be a level of simplification beyond which the model is no longer useful. It may turn out that past that point, the model has lost so much of the resolution of the system that it was designed to represent that it actually starts generating problems when we try to use it.

> A model is a representation of a real-world system and is usually created to allow humans to interact with or manage that system. If the model doesn't have the required level of detail, it may impede or impair the management of the real-world system. The level of detail required may change depending on the circumstances.

School children who begin to learn physics are told that electrons orbit the nucleus of atoms much as planets orbit a star. While this simplified model might give students an insight into the subatomic arrangement of an atom, it is a poor representation of reality. A professional physicist would be at a loss to make any progress in her field if she worked with this model of subatomic structure rather than accepting the messier but more accurate reality of probability density functions and the Pauli exclusion principle. However, what is significant is that one model can become a stepping stone to another, better model, albeit imperfect.

> The value of a model lies in whether it contains enough resolution to inform the user and whether it can represent that level of resolution in a manner that allows the user to access a more complex model (or less complex model) should the need arise.

Technical manuals, particularly electronic versions, often allow the user to access a subject with varying levels of detail. Flight Crew Operations Manuals for modern aircraft recognize that pilots do not need to know all the nuts and bolts of a particular system when attempting to resolve a problem. However, giving the user control over the depth of detail on display enables the pilot to access more detailed system descriptions if the more straightforward descriptions fail to provide sufficient guidance.

These examples demonstrate that simplified knowledge (or modeling) is only useful in so far as the user is aware of the extent of simplification and can access knowledge/models with higher or lower resolution should the need arise. One might argue that the user should always be presented with the most comprehensive description of a system to allow them to make the best possible decisions. Theoretically, this makes sense, but the price of that is the cognitive burden imposed on the user.

2.1.1.2 Forced simplification

For all our amazing mental capabilities as human beings, the actual operating system of our brain is surprisingly limited. In a later chapter on neurobiology, we will go into more detail on the mechanics of decision-making, but, suffice it to say, we are not as smart as we think we are. The process of focusing attention on something (like a model) and using the sensory information to form a perception is a time-consuming and cognitively demanding one. The greater the detail of the model, the more information must be gathered from it and folded into our perceptual model. Not only that, but we may also have to manipulate that mental model to be able to project the workings of that model into the future. The more complex the model (or our demands of the model), the higher the task load for the brain. The more pressure we are under to extract useful information, the shorter the time available for cognition. This combination of task load and time constraint when it comes to cognition determines the workload for the brain.

> Cognitive workload is a factor of task load (number of cognitive tasks combined with the complexity of those tasks) and processing time available. Workload is a critical determinant in the accuracy of cognition. If the workload is too high, perceptual models and decision-making processes are simplified to try and complete the tasks in the time available. It is this forced simplification that can lead to failure.

Under sufficient workload, the brain will simplify models and analytic processes automatically. In dynamic environments, this can easily lead to system failure (as we will see in the case studies in Part 4). Workload is a critical determinant of the accuracy of cognition.

2.1.1.3 Entrenched simplification

As laid out in the first part of this chapter, the concept of simplification suggests that we have either chosen to simplify something to make it useable or have been forced to simplify because of increasing workload. There is, however, a third and, arguably, more natural type of simplification; the model or representation seems fit for purpose, and we are not aware that reality is, in fact, more complex. To illustrate this, let's consider the field of industrial safety.

Since the early part of the 20th century, scientists in the field of industrial safety have tried to understand the causation of accidents in the workplace. In 1931, Herbert W. Heinrich published a paper entitled "Industrial Accident

Prevention: a scientific approach."[2] His conclusion was a simple one: most accidents do not occur spontaneously. He attempted to describe the process using the visual metaphor of a line of dominos. A problem in the social environment results in some manner of deficiency in a worker, which causes them to perform an unsafe act, leading to an accident and, finally, an injury. Heinrich's model was a step forward because it recognized that multiple contributing factors often led to workplace accidents.

This model persisted until the late part of the 20th century primarily because it felt like the recognition that multiple contributing factors were usually involved in accidents or incidents gave managers working in industry a way of managing the risks of workplace accidents in a manner that was better than before. Rather than simply focusing on the accidents or potential accidents themselves, they identified features of the social environment in a workplace or particular acts that workers might undertake to identify risk and reduce the probability of accident and injury. This model of accident causation became entrenched primarily because it seemed suitable to the task of managing safety.

In the latter part of the 20th century, researchers in industrial safety reexamined this model. Despite recognizing that multiple contributory factors led to workplace accidents and injuries, this model was fundamentally active: something would be wrong in the social environment, which then actively led to some manner of deficiency in a worker, which then caused them to perform an unsafe act, which ultimately led to accident and injury. Like the visual metaphor of a row of dominoes, the process was entirely active; if the first domino doesn't fall, the accident is prevented. The Manchester-based psychologist Professor James Reason developed a new model which combined the idea that accidents do not occur in isolation with the new concept of accident causation being a passive process.[3] Rather than a row of dominoes falling over, he recognized that accidents occur passively as a sequence unfolds and passes through holes in the barriers designed to prevent such progression. Fig. 2.1.1 illustrates this model:

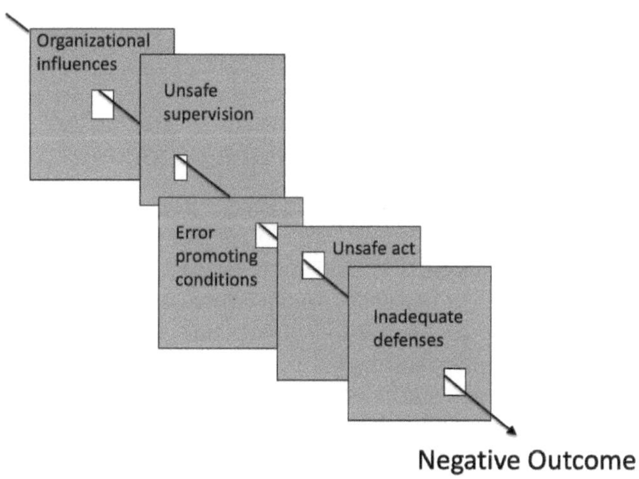

FIGURE 2.1.1 James Reason's model of accident causation.

As you can see in this model, the process is no longer active. While it is still linear, there is now a path through weakened defenses, and this process is a passive one. Although it cannot be represented in a static diagram, James Reason's initial idea was that the holes in the barriers keep changing position. As such, in one instance, the sequence of accident progression may be stopped at the second barrier, whereas in another instance, the hole has moved to allow the sequence of causation to breach that barrier before moving on to the next one.

Reason's model, sometimes known as the Swiss Cheese Model in reference to the fact that the different layers have holes in them, marked a huge step forward in our understanding of industrial safety. Across multiple industries, training and safety measures were put in place based on this model. The processes of risk management and accident investigation could now focus specifically on where holes in defensive barriers had occurred which had allowed an accident sequence to unfold. By identifying where the holes in the defense barriers were, it was now possible to implement specific measures to minimize the risks of those holes occurring in the same place again. By plugging the holes, it was possible to reduce the probability that a sequence of events could breach successive barriers and cause a negative outcome.

What the Domino Model and the Swiss Cheese Model have in common is that they are easy to visualize; a row of dominoes falling over or a set of barriers with holes in them are both beguiling visual metaphors which we think give us an insight into industrial safety and accident causation. There is also a certain intellectual satisfaction at being able to visualize the contributory factors to such ephemeral phenomena as safety and risk. In this way, we become entrenched in the belief that these models are correct and are up to the task of ensuring industrial safety. Once more, however, the reality is more complex than we might expect.

> The apparent completeness or ease of visualization of a model does not mean it is a sufficiently accurate representation of reality.

2.1.2 Necessary complexity and systems thinking

We are now at the point where we need to accept that a simple explanation is not always the best one. There will often be a necessary complexity to our models of reality that may make them unwieldy to use but will ultimately mean they have greater utility. The simplification imperative of humans is ultimately one based on deconstruction: it is easier to understand component parts than to try and understand how they interact in complex, dynamic environments. The modern world we have constructed for ourselves no longer affords us this luxury. There reaches a point where our models need to embody some of the complexity of the systems that they are trying to represent. We need to fight the human urge to deconstruct systems into component parts and instead embrace the idea that we have to think on the scale of the larger systems that these components are a part of. This approach is referred to as Systems Thinking. To illustrate this, let us continue describing the evolution of systems thinking in industrial safety.

James Reason's work showed explicitly that aside from the act itself, there are generally multiple contributory, often organizational factors involved in industrial safety (and, equally, safety failures). A further refinement of this is to consider the nature of the organizations within which we work. Consider an airline. An airline is a collection of humans, machines, and procedures that come together to achieve a purpose—the safe, efficient, and profitable transport of people and/or cargo from one point to another. In this way, we can consider an airline to be a system.

To complicate matters slightly, the airline system is composed of smaller subsystems, each of which can be considered as a system in its own right. For example, two pilots operating a flight constitute a system. The cabin crew operating that flight also constitutes a system. These two systems come together to form another system—the crew. There may be many of these crew systems operating simultaneously in an airline with multiple aircraft. These flights form part of the airline's network, itself a system, and the airline's network is part of the global transport system. Like Russian Dolls, large systems will contain multiple smaller systems. The terminology can be confusing, but in later chapters, we will look at how something that we would consider as a system when we look at it on one scale could also be regarded as a subsystem when we look at it on a different scale.

A large airline may have thousands of staff, hundreds of aircraft, and millions of lines of procedures, laws, and computer code that are required to come together to achieve the desired objective. How can we possibly predict where failure will occur in this highly complex and interconnected network? The simple answer is that we probably can't. The sheer number of components and the number of states that each component can be in mean that we cannot possibly consider every possible combination of components and component states. Consider yourself—when you are at work, you may be performing well, above-average, average, below average, or terribly. There are, of course, lots of states in between these. Still, this example is just meant to illustrate that most system components operate on a spectrum that starts at being nonfunctional and goes all the way up to maximum functionality. When we consider this, we can see that accurately predicting how such a system could fail is truly impossible as there are too many variables. We cannot visualize dominos or slices of Swiss cheese for every part of this system. We might be able to see these things retrospectively once an accident has occurred, but that doesn't help us when we are proactively trying to manage risk within the system. This is a crucial point to emphasize:

> A system failure is far easier to analyze retrospectively than predict proactively. When organizations undertake investigations, it is essential to remember that even though the chain of events may look glaringly obvious when they are analyzed retrospectively, it likely would have been impossible for the participants involved in the event to have been able to predict the complete sequence of events as they unfolded over time.

How then can we get a handle on managing such a complex system? As we will see later when we look at complex systems in more detail, to be able to manage them we have to accept that because of their complex, interconnected nature, it is impossible to predict how failures will evolve and spread throughout them and how the performance of other components could potentially cancel out or, conversely, amplify these effects. There are so many variables that we are almost in the realm of chaos theory, i.e., the idea that very small disturbances in a system can either disappear or spread causing surprisingly large consequences that are not predictable. The classic example attributed to one of the founders of chaos theory (Edward Lorenz, a meteorologist) is that a single beat of a butterfly's wing in Brazil could lead to a hurricane in Texas seven weeks later.[4]

A single, seemingly inconsequential event can determine whether a catastrophe happens or doesn't. In the same way, a single, seemingly inconsequential event could be the tipping point for a system leading to its collapse or, conversely, could prevent a catastrophe from happening in a way that we might be completely unaware of.

Here we have an excellent example of similar complex dynamics occurring in different disciplines. Small perturbations occurring within an organization can amplify until the result is many orders of magnitude greater than the initial event. Small perturbations occurring within a meteorological system can also be amplified to the extent that they become natural disasters. It would be foolish to ignore the differences between these two systems (one being a sociologically based system, the other being physical in nature). Yet, we can recognize that the *dynamics* at work are very similar.

> Complex dynamics in one type of system can give insight into the dynamics of other systems even though the nature and components of those systems may be very different. Referring back to the introduction of this book, these observations hint at the consilient nature of complex dynamics across different areas of reality.

This recognition is at the heart of what this book is trying to achieve. Systems thinking is not necessarily about deriving mathematically complete proofs and formulae that can be applied across disciplines to predict the dynamics of complex systems. Instead, it is about identifying, even anecdotally, similarities between the dynamics of different complex systems in different parts of reality that may ultimately give us insights that can help us understand and manage the systems within our own sphere of influence. In essence, this is not the search for a theory of everything, but rather an attempt to peek out of the tiny windows of the silos within which we find ourselves because of the increasingly specialized way in which we are educated and trained, to see if there are tools being deployed in other fields which may be useful to us.

Chapter 2.2

The language of systems

Essentially, all models are wrong, but some are useful.

George E. P. Box, statistician

Don't do anything until you know the result of your last action.

Warren McCulloch, cybernetician

Chapter outline

2.2.1 Interlocking systems of reality

You spend your life surrounded by systems. You interact with them, you build them, you adapt them, and, from time to time, you disrupt them. Some people have even had the distinction of having made systems collapse completely. Some of these ill-fated systems were small and relatively unimportant. Some of them were huge, and their failure may have led to massive financial losses, the collapse of a company, or even thousands of deaths. The unfortunate thing about these systems that surround you is that they rarely consist of simple, linear interactions between components, much as you would see in the cogs of a gearbox: linear, understandable, solid metal structures interacting with other linear, understandable, solid metal structures in a way that is entirely predictable and almost totally

Designing and Managing Complex Systems. https://doi.org/10.1016/B978-0-323-91609-7.00003-2

reliable. If you were to imagine the cogs in all the systems that surround you, imagine them as complicated, multifaceted three-dimensional moving parts that need to somehow mesh with other equally complicated moving parts. Hopefully, for the most part, the faces of these structures manage to align themselves well enough to be able to fulfill their function. Still, sometimes they miss, or meet at the wrong angle, or become so damaged from repeated misaligned impacts and poor maintenance that they fracture. The question now becomes, can this system still work? It is tempting to think that the systems that operate our world are clear and understandable, whereas, in the words of a renowned researcher in the field of complex systems, we should consider them as "tangled, layered networks."[1]

The multiplicity and complexity of the systems that surround us in the 21st century are a testament to the industrialized nature of our world. Prehistoric woman would not have had to contend with such entities because she had to adapt her behavior to fit in with the world, whereas we can adapt the world to fit in with our behavior. Systems also have the effect of binding us together in ways that might not seem obvious. The processes of providing electrification and sanitation to a region require a concerted effort, and the systems that were created to achieve this changed the fates of large populations through their impact on housing and industry.

In 1792, the first "modern" financial market was created in New York.[2] Twenty-four brokers came together to create a system to allow for the exchange of paper stocks in a regulated fashion. In itself, this new market was a metasystem—a system that was created to regulate the activity of other systems (those related to the creation, issuance, sale, and purchase of paper stock certificates from a particular company). The other feature of this new financial market was its capacity for evolution. From these simple beginnings, the financial market which exists in that same geographical location is exponentially faster, more complex, and, arguably, more fragile than its progenitor. That location is Wall Street.

Without eschewing 21st-century life altogether, it is impossible to exist in this world without being an operator or, at least, a part of many different systems throughout each day. These systems are often hidden but far-reaching in extent both spatially and temporally. They can be created with a goal in mind, but through evolution and changing conditions find themselves fulfilling other functions or buckling under the pressure. They can limit our capabilities as humans and also enhance them. Some would argue that they detract from some of the very aspects of our nature that define what it is to be human. However, without them, the industrialized world we live in would be inconceivable. The linking up of disparate entities in an organized fashion allows the transmission of matter, energy, and information in a highly efficient manner. High efficiency decreases costs. Decreased cost leads to increased disposable income, and this disposable income drives our modern economy.

2.2.2 Structure and function of complex systems

It is impossible to talk about systems without first giving you an overview of the language used. It is an inevitable feature of a book devoted to systems that the word "system" is ubiquitous throughout the text. Before delving too deep into the subject, it is worth laying out a few principles.

1. A system is an organized set of components that are linked up to allow them to fulfill a function.
2. The function is what the system does, i.e., its purpose.
3. The components may be arranged into different subsystems, and the subsystems together go to form the system.
4. Systems can be recursive. A subsystem may have a similar layout to the larger systems it is a part of.
5. A component is the simplest part of a system that can be given a single function. For example, a machine may fulfill a function within a system, but a part of that machine, although it fulfills a function for the machine, hasn't actually been given that function by the designers of the overall system.
6. At the most basic level of the system, the components can be:
   a. Humans
   b. Procedures (including software, rules, norms, and laws)
   c. Machines
7. While these three classes of components are often systems within themselves, for the purposes of this exploration of the nature of systems, it won't often be necessary to look at this characteristic of components any more closely.
8. A component can be a part of several systems or subsystems.
9. A metasystem is necessary to coordinate activity between systems and subsystems.
10. A closed system tries not to change its activity in the event of changes within its environment (e.g., a mechanical watch). It doesn't rely on exchanging matter, energy, or information with its environment once it has been initiated.
11. An open system varies its activity based on changes within its environment (e.g., the financial markets). It exchanges matter, energy, and information with its environment.
12. An ultrastable system attempts to maintain equilibrium in the event of disturbances.
13. Systems, subsystems, and components have a function that is assigned.
14. Systems, subsystems, and components also need to have the capability to fulfill those functions.
15. When there is a mismatch between function and capability, a system, subsystem, or component may not perform adequately.
16. The capability of components may change both independently and dependently of the activity of the rest of the system.
17. The dynamics of the system (its natural behavior) will determine how it will perform in the face of changes within the environment and when reacting to the variable performance of the subsystems and components of which it is made.

2.2.3 Natural dynamics of complex systems

Some systems organize themselves spontaneously. Some are designed. Those that organize themselves spontaneously and are unconstrained by the intent of some higher power in their evolution will tend toward efficiency through the natural drives to minimize the expenditure of energy. The human body is an excellent example of this. Through the process of natural selection, we have evolved in structure and function to be able to exist successfully in the world. Unfortunately, because of the slow pace of evolution, the world we have evolved to fit into so successfully was that of our prehistoric ancestors. Once we reached a certain point in our evolution, we were able to shape the world in which we lived, and it is only in the past 200 years that this ability to shape the world has accelerated exponentially. Humans have reached a point where our ability to create our world has outstripped our cognitive capability to keep up with what we are creating. Two hundred years is far too short to see any evolutionary changes through natural selection within our species. Still, the 21st century will likely see humans evolve by means other than natural selection.

Advances in electrical engineering can allow humans to augment themselves with technology to better interface with the computer systems that increasingly dictate the shape of the world around us. More significantly, advances in genetic engineering have brought our species to the point where we can actually alter the structure of our DNA to add, subtract, or change our characteristics, and we can do so in such a way that these changes are inheritable. For the first time in history, humans can direct the course of their own evolution. Sensibly, the deployment of these technologies is closely monitored, and the most extreme interventions, particularly those which can affect the human germline (i.e., introduce inheritable genetic changes), are prohibited.[3]

So, in the case of humans, our biological complexity came about through evolutionary processes which drove us from simple monocellular organisms to complex multicellular animals. This spontaneous development has got us to the point where our future evolution may be by our own design, particularly in the short term. In a later chapter, we will look more closely at the processes which drive spontaneous, self-organizing systems, but for now, it is enough to state the following:

Complex systems may evolve spontaneously, or they can be created by design.

2.2.4 The need for complex systems

Ross Ashby, one of the founders of the cybernetic movement, realized a fundamental relationship between systems and the environments in which they operate. To deal with a wide variety of possible environmental conditions, the system must be able to have at least as many states as the operating environment can be in, accounting for possible disturbances and pressures. More formally, he stated:

> When the variety or complexity of the environment exceeds the capacity of a system (natural or artificial), the environment will dominate and ultimately destroy that system.[4]

Variety is a term that denotes how many states a system or the environment can take. For a system to manage the amount of variety that is possible in the environment, it must be able to exhibit at least the same degree of variety or, preferably, even more.

Looking at this statement in a slightly different way may help demonstrate the relevance of Ashby's Law a bit more clearly:

> The more we expect our systems to do and the more resilient we expect them to be in the face of disturbances exerted by the dynamic environment in which they function, the more complex the system has to become.

If an organization faces a new set of governmental regulations that might impact how it operates, it needs to account for these in how it operates. Subsystems may need to be created within the organization to check for new regulations and ensure that the business is operating in line with the current ones. The environmental complexity has increased and so must the system complexity.

Similarly, a "disturbance" to the system may come from within. An organization that decides to institute a policy of cost-cutting, bureaucratic streamlining, and a ruthless campaign against waste must first organize a group to manage this task. This group needs to be resourced and its activities coordinated throughout the rest of the organization. Creating a subsystem (the group) requires a metasystem to allow coordination with the larger system. In the short term, at least, the complexity of the organization increases. However, this subsystem and its metasystem may be erased when the task is complete, much in the same way as the installation process associated with a newly downloaded computer program often finishes the process by deleting the installation software.

> The more complex the world becomes, the more complex our systems need to be to cope with it.

2.2.5 Tractability and effective complexity

We have reached the point in the book where we need to delve further into the concept of complexity. I have managed to avoid this task so far because the idea of complexity seems like a fairly intuitive one. We seem to have an innate sense of when something is complex and when something is simple. However, our perception of something's complexity or simplicity may differ from another person's. Specific questions inevitably arise when we try to get a deeper understanding of this issue. Is there an objective way of classifying complexity? When does a system become complex? As a starting point, I will address the question, "why should we care?".

Systems come in all shapes and sizes. A better way of stating that might be to say that systems vary across their dimensions. These dimensions may be spatial, temporal, or both. To further complicate the picture, systems vary in relation to the nature and number of their components. Some systems might contain mainly human components, and some may contain mainly mechanical components. How closely related components are to each other would be another measure relevant to a particular system, as would how understandable the arrangement of the system is. If we tried to classify systems based on their differing measures across these many dimensions, the resulting diagram (if such a thing were possible given the number of potential dimensions) would be a mess.

The reason why we should care about this is at the heart of this book. In short, similarly structured systems tend to behave in similar ways. If we can understand these dynamics, we have a better chance of successfully designing and managing such systems? The problem with trying to classify different groups of systems is their sheer variability. How does one compare a computer and a factory? How does the global financial market relate to the organization of a hospital? Does an assembly line have anything in common with a school? To try and make some sense of the diverse plethora of systems that are functioning in the world, we somehow need to find a way of dividing them up in a manner that is workable, understandable, and, most crucially, suggests the strategies we should employ to best design and manage them.

Different approaches have been taken over the years, some of which led to fruitful discoveries and some of which seemed to do nothing but muddy the waters. The development of General Systems Theory in the middle part of the 20th century suggested some interesting features that different systems may have in addition to the principles we examined earlier[5]:

- The notion of subsystems and components within mechanical, biological, and social systems.

- Open systems – those that exchange information, energy, or material with their surroundings.
- System boundaries – separate the system from its surroundings. Less well defined in open systems.
- Multiple-goal seeking – different components/subsystems within the system may have different goals.
- Equifinality – biological and social systems operate differently from mechanical ones because system endpoints can be achieved from various starting conditions, i.e., two systems exhibiting equifinality can reach the same endpoint from different starting conditions.

While some of these descriptions can be useful in describing systems, they are less useful in *defining* systems. How best then should we define systems? Do we do it with reference to the nature of their components (humans, machines, etc.)? Is it best to talk about scale, both temporal and spatial (local, national, transnational, cultural, etc.)? Do we try and pigeonhole systems into whatever academic discipline we think is best suited to study them (physics, sociology, economics, aesthetics, etc.)? All of these approaches potentially have their merits, but all share the downside that we will inevitably begin to define systems according to the silo that we think they best fit into. Mechanical systems will be analyzed from the perspective of a mechanical engineer; sociological systems will be examined from the perspective of a sociologist; and so on.

In 1984, the sociologist Charles Perrow wrote a book called "Normal Accidents; Living with high risk technologies."[6] Fresh from analyzing what had occurred during the Three Mile Island nuclear disaster in 1978, Perrow had the insight that the modern industrial systems containing people, machines, and procedures (also known as "sociotechnical systems") had become so complex that failure and success were no longer predictable. Perrow was the first person to appreciate that because of the complex, interconnected nature of the systems within which we work, it was becoming impossible to predict how failures would evolve and spread throughout the system and how the performance of other subsystems and components could potentially cancel out or, conversely, amplify these effects. In essence, Perrow thought that failure and success were two sides of the same coin.

He went further and attempted to classify complex sociotechnical systems in terms of two basic characteristics:

1. Coupling – This is how quickly matter, energy, information, or, indeed, disruption in one part of the system spreads to other parts of the system.

2. Interactions – These range from linear to complex and describe how understandable the system is to an external observer. For example, the workings of a dam are easier to imagine than the workings of a nuclear power plant. This term "interactions" is sometimes replaced with the term "tractability," i.e., understandability.[7] An intractable system is one where the observer cannot completely conceive all the components that allow the system to function, or by the time the observer has comprehended them, the system state has changed.

Perrow devised a two-axis coordinate system to try and classify different types of sociotechnical systems. For example, simple manufacturing processes would be highly tractable but loosely coupled, i.e., the manufacturing process was understandable, but a disruption in one part of that system wouldn't necessarily propagate quickly throughout the rest of the system. Power grids were moderately tractable (moderately understandable) but tightly coupled. A failure in one part of the grid would quickly have a knock-on effect in other parts of the grid.

As a paradigm for a complex sociotechnical system that is highly intractable and highly coupled, Perrow selected financial systems, and the Global Financial Crisis of 2008 bore out the accuracy of his analysis.

Given that the nature and aim of this book are to provide insights into the design and management of systems, particularly complex ones, by drawing on different academic disciplines, tractability can be seen as essential in defining different types of systems.

Some systems are inherently simple. A small system made of just a few components, for example, a balance type weighing scale, is readily understandable. While still being purely mechanical, a motor car is understandable once someone understands the mechanics, although it can clearly be seen as a more complicated system than the weighing scales. A modern nuclear reactor is still a mechanical system at heart but is even more complicated.

Note how these three systems are all mechanical systems. For the moment, we are ignoring the messiness that inevitably ensues when humans are added to the system. However, as a first step, these different types of mechanical systems illustrate a point that will be important when we later describe how we can make sense of different types of systems. This attempt at classifying these different types of mechanical systems relies on one crucial variable: the knowledge state of the observer. The workings of a nuclear reactor should be more straightforward to understand for a nuclear engineer than a car mechanic.

> Methods for making sense of systems need to account for the knowledge state of the observer.

2.2.5.1 The observer effect

Before we delve any deeper into how we make sense of different types of systems it is worth pausing to consider how the role of the observer is essential to this process. We necessarily make sense of the world and the systems which comprise it based on our knowledge and experience. The significance of the observer and how she relates to what is being observed is already an accepted phenomenon in physics but less so with regard to systems thinking.

In the early part of the 20th century, when quantum physics was beginning to overtake classical physics as the most reliable way of understanding matter and energy in the universe, the German physicist Werner Heisenberg was grappling with the behavior of electrons. At school, students learn that electrons orbit the nucleus of an atom in one of several (depending on the atom) neatly arranged levels known as shells. This planetary model of the atom, the electrons representing different planets orbiting at different distances from the sun, is sufficient to allow students to understand the basic principles of atomic structure. Still, it is a gross oversimplification when compared with reality. Electrons don't orbit nuclei in prescribed tracks but exist as a mist of particles traveling at about 1% the speed of light. Their location isn't entirely random, and the concept of shells actually relates to the probability that a particular electron will be found within a particular volume of the atom at a particular time. Electrons that might be described to a school student as being in the outer shell orbiting the nucleus are actually just more *likely* to be found in that region that is more distant from the nucleus. As if quantum physics wasn't muddying the clear waters promised by classical physics enough, electrons, in common with many other particles, exist both as particles and waves simultaneously. Heisenberg's insight was that if you could accurately describe an electron's location, you had to do this by viewing it as a particle. In this case, the electron's wave characteristics (its wavelength and momentum) couldn't be ascertained. Heisenberg called this the Uncertainty Principle, and it suggests that we can only ascertain a limited amount of information about a particular particle at a specific time.

The observer effect is a related but different phenomenon. This effect essentially says that the act of observing something changes it. Consider shining a light onto a leaf to examine it. The photons of light will cause some changes, however minor, in the leaf that would not have existed had you not illuminated it. This effect becomes more pronounced at smaller scales. To observe subatomic particles, some form of energy needs to be used. The photons from the energy source, although essentially massless, do have enough mass to change the position and/or velocity of the particles they interact with.

Both the observer effect and the uncertainty principle bring to light an essential factor in the act of looking at systems, whether it be systems on the subatomic scale or at the scale of organizations: the observer.

The significance of the observer when trying to make sense of different types of systems has traditionally been understated. Almost everyone standing in front of a light switch with even the most rudimentary grasp of electricity can understand it. However, most people standing in front of a magnetic resonance imaging (MRI) machine would be baffled by its mechanics. Does this mean that the MRI system is complicated? What happens if the observer is the engineer who designed it? Has the system now become simple?

The role of the observer in being able to tease out the nature of the systems that they are observing or interacting with is one of the features of a framework that gives insight into this concept and which we will cover shortly.

> The act of observing effects that which is being observed.

The other related observer effect is that the view changes depending on how close you are to what you are looking at. This may seem an obvious comment when it comes to observing something visually, but it becomes a bit more nuanced when we start thinking about how we view systems. Our view on the relative complexity or simplicity of a system will be affected by how closely we look at it. Selecting an appropriate scale is vital when managing a system and making decisions about it. Scales will be a fundamental concept to consider throughout this book.

> The complexity of a system will seem to change depending on who is looking at it and how closely they are looking.

2.2.6 Cynefin framework

The management consultant Dave Snowden developed the Cynefin ("cu-nev-in") framework in 1999. It has since been published in the Harvard Business Review and deployed in various fields. The Harvard Business Review article is excellent and is well worth reading.[8] Before looking at it in more detail, we just need to clarify some terminology related to the different functional areas of the framework. The original framework referred to these areas as being different "contexts." To fit in better with some of the terminology used later in the book, I refer to these different areas as "domains."

In the introduction, I said that this model was one of the main concepts I wanted to introduce in this book. It's an excellent way of understanding how different types of systems (or different operating states of the same system) require a different way of thinking, and it gives us some insight into how we can use this information to exert some measure of control over them. The model is shown in Fig. 2.2.1:

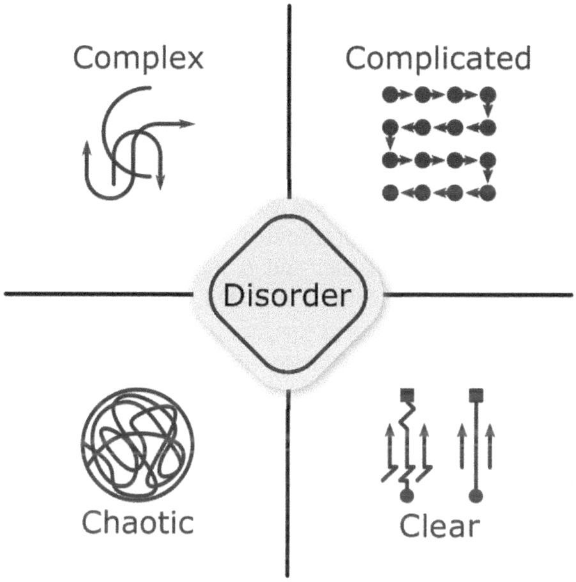

FIGURE 2.2.1 The Cynefin framework. *Image copyright Graphic Grid (www.shutterstock.com).*

In his description of this framework, Dave Snowden stresses that it is not designed as a way to categorize systems but rather an approach that allows people to make sense of what they are seeing and have a better chance of applying the appropriate management strategies.

Different types of systems may more naturally tend toward being in one domain rather than another (e.g., purely technical systems are more likely to be in the Clear or Complicated domain). It is also important to clarify that just because a system tends to operate in one particular domain, it doesn't mean that it will always stay in that domain. The same system may shift from one domain to another depending on what is actually happening to it. A severe disturbance may cause a domain shift from the Complicated domain to the Complex domain. It may later shift back into the Complicated domain depending on how successful the management strategies employed actually were.

This idea of domain shift will become more critical as the book goes on because it can explain why managers need to be ready to adjust their decision-making style if such domain shifts do occur.

We also need to consider what scale we are working at. Much like the way that a change to an observer's position relative to that which is being observed is going to give a different view (potentially a very different view when going from macroscopic to microscopic scales), so too can our assessment of what domain a system is operating in change depending on how we view it. A factory that uses production lines may seem like a Clear system when taking a system-wide view: raw material comes in and is converted into something else that is then sold and distributed. However, a management decision may be about a subsystem that is operating in the Complex domain, e.g., a human resources issue involving multiple staff members, new legislation, and workers' union rights. A management strategy based on Clear domain thinking is unlikely to be successful, so we need to be conscious of selecting the correct scale to view the problem. Domains of one type can be embedded in domains of another type and might only be revealed when viewed on a different scale.

As this can be a difficult issue, and it might be difficult to completely visualize how domains and systems are related, here is a summary of this last section. It is essential to understand the fluidity that is needed when trying to work out what domain is relevant at any one time:

1. Certain types of systems may naturally tend to operate in a particular domain
2. Just because a system naturally operates in one domain doesn't mean it won't shift to another
3. These domain shifts mean different types of responses are required
4. We may encounter subsystems operating in different domain types to the overall system as we zoom in

As this framework will be so important for what follows, it will be time well spent going into it in more detail. There are actually five domains in the model, and we will look at the characteristics of each one. To best understand this framework, imagine that you are managing a system (or a problem in a system) that is operating in one of these domains. Your approach will need to be different depending on what domain you think you are dealing with.

2.2.6.1 Clear domain

In this domain, effect clearly follows cause. Patterns repeat themselves, and the outcomes are usually the same. When trying to manage a system that behaves in this manner, the approach is sense-categorize-respond; sense the current situation, categorize it according to what you know, and respond with the solution that procedure suggests you apply. In this way, a system operating in the Clear domain responds to best practice. Examples of systems that would mainly operate in the Clear domain would be clearly defined manufacturing or industrial processes, i.e., sociotechnical systems that are predominantly technical and where the machines involved are readily understandable (tractable).

2.2.6.2 Complicated domain

In this domain, effect follows cause in a way that may not be obvious or may be masked by multiple other effects. However, these relationships are discoverable. The Complicated domain can be viewed as the domain of expertise. The management response is sense-analyze-respond; sense the current situation, analyze the nature of the problem, respond by deploying experts with experience of that particular problem. There is no obvious best practice, only good practice driven by expertise. Examples of systems that would mainly operate in the Complicated domain would again be socio-technical in nature but where the underlying operation of the machines may not always be readily apparent to the humans involved. For example, a highly automated nuclear power plant.

2.2.6.3 Complex domain

In this domain, effect does not clearly follow cause. Changes ripple quickly through the system in a manner that may mask the initial events. It will not be obvious which approach is required to manage a system in the Complex domain. Because it isn't possible to derive a fact-based response, the management response needs to be probe-sense-respond; probe the system by trying different management strategies, sense which of them confers the best degree of control, and respond by scaling up this strategy. If the response doesn't work, a different response can be tested. It may be that testing a particular response worsens the situation. Because of this, managers need to be sure that they have a way of dampening the effects of a test so that it doesn't cause the situation to deteriorate. Adequate practice should develop over time and, possibly, after multiple tests building on each other, so we call this iterative practice. (In the Cynefin framework, this is referred to as emergent practice but, for the purposes of this book, it will be clearer later on if we refer to it as iterative practice).

Examples of systems that would mainly operate in the Complex domain would be those where the workings of the system depend more heavily on the interrelated activities of multiple component types, and especially when humans are heavily involved in the running of the system. For example, a less automated nuclear power plant that relied on human operators to oversee and intervene in essential plant functions. Other examples would include financial institutions or transportation systems.

2.2.6.4 Chaotic domain

With a system operating in this domain (or, more likely, not operating), there is no link between cause and effect. The system is changing unpredictably and often more quickly than a human operator can keep up with. The system is unconstrained and ungovernable, with the potential for catastrophic failure. The management strategy is act-sense-respond; act instinctively to prevent system failure, sense how that action has affected the system, and respond by trying to shift the system into one of the other domains. There is no time for careful probing as we might do in the complex domain, and so we refer to this as novel practice. An example would be the immediate situation after a large natural disaster.

2.2.6.5 Disorder

To deploy the correct management strategy for the domain that the system is in, a manager needs to know *which* domain the system is in. The accuracy of this determination depends on the knowledge state of the manager and their view of the issue (i.e., the scale at which they are analyzing it). We have to accept that not everyone will know what domain a system is in at all times. This is reflected in the framework diagram by the central area called "disorder." It is important to note that in this description, "disorder" does not necessarily refer to the state of the system but instead it reflects the fact that the person managing the system cannot determine with any certainty what domain the system is in and, hence, won't know the appropriate management strategy to use. It may be that a domain shift is occurring or that some parts of the situation exist in one domain and some parts in another. Snowden recommends breaking the situation down into smaller parts and then attempting to determine which domain those parts best fit into.

2.2.6.6 The clear/chaotic boundary

As you might have worked out already, when a domain shift occurs, generally because of a disturbance, the most likely direction of this shift is counterclockwise. For example, a disturbance affecting a system in the Complicated domain may shift it into the Complex domain. A disturbance affecting a system in the Complex domain may shift it into the Chaotic domain. These boundaries may be subtle, and it might not be immediately clear that a domain shift has occurred. The one boundary that is different is between the Clear and Chaotic domains. In most cases, a system will gradually transition between domains except in the case of transition from Clear to Chaotic. This has been likened to falling off a cliff in that it happens very quickly, and the outcome is usually disastrous. It is generally reflected on representations of the Cynefin framework as a cliff edge, the high point being on the Clear side of the boundary. Danger signs that this serious domain shift might occur include increasing complacency, persistent oversimplification of problems, and an overreliance on trying to find best practice solutions to problems that aren't amenable to that form of management. An example of this sort of transition would be if a highly mechanized and ordinarily stable system becomes overreliant on control via automated systems. Should the control software for the automation not be updated and secured appropriately, a relatively minor failure could spread throughout the system and drive it into the Chaotic domain.

2.2.7 How humans affect complexity

The advantage of the Cynefin framework as a structure on which to build our understanding of complex systems is that it recognizes that certain types of systems will naturally fit into one domain rather than another but also that the same system may transition between different domains (domain shift) depending on how it reacts to external or internal disturbances. It also reflects the importance of the observer and that only with an accurate understanding of what domain a system is currently operating in will the observer (manager) be able to deploy the appropriate management strategies.

The knowledge state of the observer determines how quickly she will be able to ascertain what domain a system is operating in. For an observer who is used to working in the Clear domain, it may be difficult to accept that the system state has shifted to a different domain and that a different management strategy is required. The observer (who is also managing the system) is part of that system, so her knowledge state is relevant when determining the system's state. An expert might be brought in to manage a system that is operating in the Complicated domain. Still, should the nature of her expertise not be appropriate to the situation, it may drive the system into the Complex domain. In this way, we can see that the state of a system is intrinsically bound up with the observer/manager and their interventions.

In this book, we will frequently encounter the importance of scales. In the Cynefin framework, a system may seem to move from different domains as we decrease the scale of our observation. For example, from a distance and when being considered in the context of other systems, the underground metro system for a city may seem to be in the Clear domain: trains arrive, and trains depart according to a timetable. With that timetable, an observer would be able to make reasonable predictions about where trains will be at any particular time. This regularity would indeed suggest that this system is in the Clear domain. However, if we zoom in, we now see that things are a little bit more complicated and the signaling system that tells trains when to move and directs them onto different tracks is dependent on other trains in the vicinity. Now it looks like we are in the Complicated domain. If we zoom in even further, we can see the complex system with all its people, machines, and procedures in a highly coupled network. This isn't to say that the first two observations are invalid. We know that we necessarily need to simplify to make models of situations. If we only had to consider the metro system in the context of the power grid of the metropolitan area, modeling it working in the Clear domain might be sufficient for our purposes. The problems occur when we select an inappropriately shallow or deep analytical view which then oversimplifies or overcomplicates the process.

Finally, we must consider one feature of systems that tend toward the Complex domain: those containing humans. In the Clear domain, processes are clear, cause clearly precedes effect, and we can apply best practice without a great deal of consideration. An assembly line would be a good example. An assembly line may be purely mechanical, or it may include human components that operate in a clearly prescribed manner. It could be argued that the humans in this example are operating mechanistically, again depending on what scale we are using.

In the Complicated domain, multiple technical processes may come together to achieve an objective, and human operators oversee these technical processes. These operators may be separated spatially and temporally, so it will not always be possible for them to have complete knowledge of what other component parts of the system are doing at any one time. This potential masking of cause and effect means the system state is not readily observable at any one time but is still constrained in some way, usually by technical limits on the system's mechanical parts. A nuclear power plant would be a good example of this.

In the Complex domain, humans are usually the most significant components in the system. This may be a system that requires few mechanical components, such as the board room or management building of a large multinational corporation. Because of the sheer variability and flexibility of human performance, the operation of such a system is potentially highly variable. You may be beginning to see that it is the range of human capability and their relatively unpredictable performance compared to machines that mean that when a system becomes more heavily dependent on its human components, it is more likely to be operating in the Complex domain. At first glance, this may seem disadvantageous. Why would we want to operate in the obscure, unpredictable, Complex domain when instead we could try and get the system back into the Complicated or even Clear domain by bringing in more rules and procedures to constrain and prescribe the activities of the human components? The answer to that brings us back to the previous chapter on the need for complex systems: the Law of Requisite Variety:

> The more we expect our systems to do and the more resilient we expect them to be in the face of disturbances exerted by the dynamic environment in which they function, the more complex the system has to become.

Adding humans to the system and giving them enough scope to deploy their innate flexibility can increase the resilience of a system and increase its capability in the face of external and internal disturbances.

> More human components tend to drive a system toward the Complex domain. This can be vital when the system is designed to deal with a wide variety of conditions and disturbances as it increases the number of states the system can be in and so satisfies the Law of Requisite Variety.

Chapter 2.3

Classes of systems

At the heart there is a hollow sun by which we are constructed and undone
Don Paterson, from his poem Francesca Woodman

Chapter outline

2.3.1 Classes of systems

At first glance, this chapter may appear to be a whistle-stop tour of all natural history from the Big Bang to the present day. Given that the entire universe comprises of systems, such a history can give us an insight into the natural dynamics of such systems, given that so many of them managed to organize themselves and went on to function perfectly well long before humans arrived on the scene. The idea of consilience suggests that there is something to be learned from how the universe manages to create for itself self-managing, self-actualizing systems, and this chapter will give a flavor of that.

The description in Part 1 of this book of how the human nervous system came to be used as a model for controlling complex systems highlighted that self-evolving, self-organizing biological systems tend to be very good at their job and that we can learn much from them. The prime example is the human nervous system. Later, in this part of the book, we will go into the human nervous system

Designing and Managing Complex Systems. https://doi.org/10.1016/B978-0-323-91609-7.00004-4

in more detail and explain why it makes such a good template for the complex systems that we seek to design ourselves. However, the human nervous system did not occur spontaneously. Through the process of evolution, nature progressively created a complex system with the requisite variety to allow us, the organism, to function and even thrive on this potentially hostile planet.

In attempting to explain the origins of the human nervous system, we inevitably come up against the "stop rule." How far back do we go? Suppose we accept that a system of such complexity developed progressively over many thousands of years. In that case, we must explore the concept of spontaneous self-organization, i.e., how smaller components and subsystems assembled themselves into structures of greater complexity. Some of these steps seem to be governed by physical laws, while others could arguably be ascribed to chance. To be as complete as possible, we will start at the very beginning of the physical universe as a prelude to talking about the biological universe and how that leads us on to the complex systems of our society. As history unfolds, so different types of systems come into being based on the degree of order and complexity in the universe. The relationship between the different classes of systems is shown in Fig. 2.3.1:

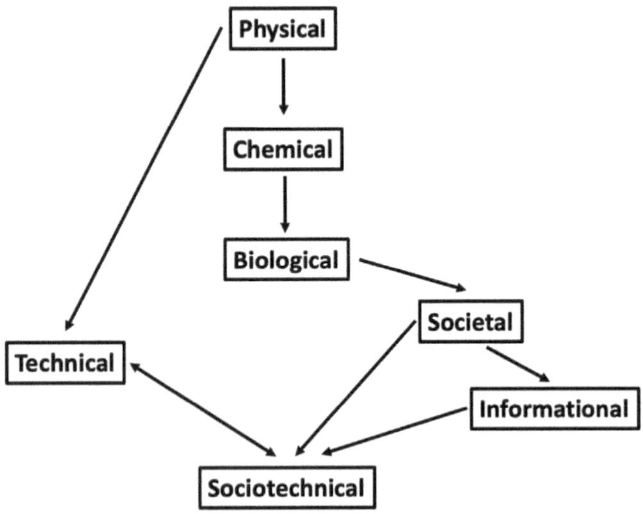

FIGURE 2.3.1 The different classes of systems.

2.3.2 Physical systems

2.3.2.1 Big Bang chemistry

When the universe was 10^{-37} seconds old (i.e., very young indeed, this being an extremely brief time after the Big Bang), something happened that will give insight into the rest of this chapter. With the universe encompassing a much smaller volume than it does now, tiny random fluctuations associated with the quantum states of particles gave this expanding volume of matter and energy a very subtle irregularity. The subsequent process of expansion of the universe meant that these tiny irregularities turned into the unimaginably massive irregularities in the structure of the universe, i.e., this is part of the reason that most galaxies form in regions with other galaxies leaving vast voids in between these conglomerations.

> Quantum mechanical indeterminacy at the level of subatomic particles can drive systems away from regularity in a way that persists as the system evolves.

From a roiling sea of subatomic particles, the first simple atomic nuclei formed when the energy state of the universe was at the correct level to permit these subatomic particles to combine. This cooling mass of protons and neutrons organized themselves into the nuclei of the three simplest elements on the periodic table: hydrogen, helium, and lithium (although lithium was a lot rarer). This process is known as Big Bang nucleosynthesis. It would take another 380,000 years for the electrons in the universe to reach the right energy state to be able to combine with these nuclei and form neutral atoms of these three simple elements.

> In some cases, spontaneous self-organization requires the components to be in a specific energy state.

As the universe expanded further, gravitation caused these simple chemical elements to aggregate together. In a process of positive feedback, the larger this aggregation of material became, the greater would be its gravitational pull and the more material it would attract. Eventually, this aggregation of matter would lead to gravity so strong that the simple elements within its mass would be forced closer and closer together at ever higher temperatures and pressures. Eventually, hydrogen atoms would become so tightly compressed that they would combine and become a helium nucleus through nuclear fusion. This fusion reaction also released a massive amount of energy, and in this manner, these aggregations of matter eventually sparked into stars such as our sun. Other distributed matter in the area would be swept into the orbit of these newly born stars and, in their turn, aggregate into planets. Interestingly, planets outside of our own solar system were thought to be rare, but recent studies of stars outside of our solar system suggest that they are far more prevalent.[1]

These stars weren't just helium factories, however. Through the process of stellar nucleosynthesis, a vast array of chemical elements was formed. Depending on the physical nature of the star in question, different elements were created. Low-mass stars that eventually fizzled out ejected many lighter elements out into the universe. Larger stars that expired by going supernova and ejecting practically all their mass in one massive explosion seeded the universe with many heavier elements. Strange, exotic stellar bodies such as neutron stars collided and merged to create the heaviest, most complicated elements, many of which were radioactive in themselves. The grinding, clockwork, physical processes of stellar birth, death, and collision eventually seeded the universe with the chemical elements that would later form more complex structures.

> Simple physical laws can increase complexity and diversity by acting in predictable ways upon simpler components.

2.3.2.2 Complexity and uncertainty at the subatomic level

It would seem natural to press on from the last section to look at how these chemical elements combined to form inorganic molecules (those that don't contain carbon compounds in certain arrangements), which, in their turn, created the organic molecules (those which always contain carbon, usually with hydrogen), which are the building blocks of life. However, we must take a big step backward before we move on. For simplicity, the previous section described every step of the process after the Big Bang until the formation of chemical elements. As with many things in complexity science, the description was meant to give you a model of the process without going into too much detail. I'm hoping that the process as described made sense but, I'm afraid I left out an essential bit of the picture because I think it deserves to be considered separately. It is also true to say that a lot of what follows was only discovered recently, but I believe it has as big an impact on complexity science as it had on physics.

As we have seen, some of the elements created in these stellar foundries have an inherently unstable structure. Again, energy states at the time dictate how these elements will behave, much as the energy state of the early universe dictated how matter formed. This instability manifests itself as radioactivity. An unstable element might emit radiation and so split into simpler elements. Conversely, other elements will absorb radiation and, in their turn, become more complex.

Physicists, particularly in the 19th and 20th centuries, attempted to understand this process by peering deeper and deeper into the structure of atoms. The word atom itself means uncuttable, although it quickly became apparent that this was a misnomer as there are deeper structures within the atom which we are progressively understanding. It seems that each time we think we have reached an understanding of the simplest building blocks of matter, we find that there is a level further down.

So far, the smallest building blocks that I have mentioned have been the protons, neutrons, and electrons, which go together to form atoms. Electrons themselves are extremely small and can be considered fundamental particles in their own right. However, protons and neutrons are considerably larger and physicists in the 20th century theorized that these particles, now known as examples of baryons, contained smaller subunits. The physicist Murray Gell-Mann first formally theorized that baryons comprised smaller structures. His first mathematical representations of the constituent parts of baryons suggested that there were three subunits within each.

Interestingly, a minor controversy broke out when some scientists suggested that Gell-Mann had only meant to suggest that these three elements were essentially mathematical in nature, i.e., a way of making certain formulae work. Subsequent findings by experimental physicists found that these baryons did indeed contain three physical subunits.[2] What was remarkable about the subunits was that there seemed to be no way of separating them. These three subunits existed in a triad within certain baryons such as protons and neutrons, which appeared to be entirely indivisible. The subunits were subsequently christened quarks.

In his biography, it was reported that Murray Gell-Mann went to a doctor who was studying at his university to get a prescription that said that discussing philosophy was bad for his health. Whenever it looked like discussions were straying into the realms of metaphysics, he could use the prescription to direct the discussion back to matters of physics.[3] For scientists probing the nature of the universe at the very edges of our comprehension, it must be tempting to sometimes tip over into philosophical musings about what sort of plan was behind the indescribably diverse reality with which we are presented. When it turned out that quarks only existed in any sort of meaningful way as a system with other quarks connected by force carriers known as gluons, he must have had to refer to his prescription more frequently.

Because of our simplification imperative, humans are programmed to believe that everything can be reduced to simpler components. From a loaf of bread, we can get a slice of bread. From a slice of bread, we can get a crumb. From a crumb, we can get the proteins and carbohydrates that make up the structure of the bread. From the carbohydrates, we can get the chemical elements. From the chemical elements, we can get the protons, neutrons, and electrons and whatever complexities they contain. To a certain extent, our reductionism tends toward simplification. However, when we arrive at the atomic and the subatomic level, the physics gets a lot weirder.

It is fascinating that most of the basic building blocks of matter in the universe are, in fact, complex systems of quarks and gluons, which have no meaningful existence separately. The most fundamental building block of the universe that we know of is not some dumb, homogeneous blob of matter, but rather a teaming, complex, interacting soup of tightly constrained components that do not seem to exist naturally in isolation outside of the baryons. There is some evidence to suggest that by applying immense energies, some baryons can be broken down and become what is known as a quark-gluon plasma.

However, this type of plasma doesn't appear to exist in nature although it may have been present shortly after the Big Bang.

As if that were not enough, when we go one level up from the baryons, we get to the larger structure of the atom. Our early schooling in the physical sciences tells us that atoms are comprised of protons, neutrons, and electrons, the electrons orbiting in well-described orbits around the central nucleus. We have already described that this is not the whole story and that the electrons exist as a sort of cloud around the nucleus. The insights of quantum physicists tell us that we cannot accurately measure both an electron's position and its velocity in any meaningful way. This is the Heisenberg Uncertainty Principle. You might be asking yourself now, why does this matter?

Consider what we covered about the radioactive elements. A standard measure of radioactivity for a given substance is its half-life. The half-life of a radioactive element is a statistical measure that tells us how long it will take for this radioactive element to decay into a nonradioactive, stable element, i.e., how long it will take for this sample of the material to lose half of its radioactivity. The interesting feature is this; it is very difficult to predict precisely when one radioactive atom will emit a tiny amount of radiation and so decay into a smaller atom. All we can do is predict the probability. It is this phenomenon that forms the basis of the famous "Schrödinger's Cat" thought experiment.

Unlike the standard Newtonian view of the universe, where the movement of planets, stars, and moons could be described with relatively simple equations, it was found that these simple equations couldn't be generalized to smaller scales. Quantum mechanics at this level is really a dance of probabilities. We can no longer deal in the absolutes of Newtonian mechanics but must instead fall back on the messier realm of quantum mechanics; some things have a probability of happening, which means that they may happen, but they may not occur in the timeframe that we expect.

Tying this back to the electron, the formation of molecules through the combination of atoms will generally involve some exchange or sharing of their electrons. The findings of quantum mechanics tell us that we can never be sure about how and when these reactions will occur. We can only speak to their probabilities.

In a book that is meant to be about the management of complex systems, you might be asking yourself why I'm spending such a long time concentrating on the inner workings of atoms. The reason is that complex systems are defined to a certain extent by uncertainty. If a system were entirely predictable, we wouldn't be able to call it complex. I need to stress that uncertainty isn't the only determining feature of complex systems, but it is nevertheless an important one. This lack of predictability needs to come from somewhere. It goes back to the old Newtonian view of the universe. An idealistic reading of classical Newtonian mechanics suggests that the universe, from the level of the planets and stars down to the level of individual atoms, should be entirely predictable and describable using standard equations. If this were the case, we could consider the universe to be one giant machine operating in a wholly predictable, understandable fashion. From the birth of the universe, we would theoretically be able to map the trajectory of every atom and every subatomic particle, describe their collisions with other atoms and subatomic particles, describe the molecules that developed from there, and essentially, describe the entire evolution of reality as we know it. We like to believe that the complexity and uncertainty in the universe come from its most complicated and "important" components, i.e., us humans. If we are unpredictable components within complex systems, it means that the functioning of the system isn't predictable either. To a certain extent, I think this is true. However, we do need to examine where *our* unpredictability comes from. Later on in this part of the book, we will look at the neurobiology of human beings, specifically, how neurons function. But suffice it to say, we should be very grateful for the uncertainty which is embodied in the simple building blocks of the universe, such as atoms, because it is this uncertainty that generates possibility. In fact, rather than thinking of atoms as boring building blocks of the universe, I believe we should imagine them as tiny little possibility generators simply because their behavior cannot be determined with certainty.

> Complexity isn't just confined to large-scale structures in the universe. Smaller-scale structures, such as subatomic particles, are also complex in their way, and atoms themselves, thanks to quantum mechanics, are not entirely predictable in how they behave, especially when it comes to interacting with other atoms. As such, the matter in the universe is actually built up of tiny possibility generators, and this contributes to the complexity of reality.

2.3.3 Biological systems

So far, we have described how Big Bang nucleosynthesis and stellar nucleosynthesis have created the diversity of atomic elements that we see in the universe. These elements combined and formed more complex molecules on the surfaces of planets and asteroids orbiting through space. The heating and cooling of these astral bodies drove chemical reactions which combined these simpler elements into more complex inorganic molecules, such as water. Collisions of planetary bodies generated vast quantities of space dust which were then seeded with elements formed through stellar nucleosynthesis. However, a step must have occurred where these inorganic molecules went on to self-organize into organic molecules. What do we mean by organic molecules? Essentially, organic molecules are those that constitute living organisms and are based on carbon. All life on Earth is carbon-based.

Interestingly, given that carbon is only one of over 100 elements in the periodic table, there seems to be far more carbon in the universe than would ordinarily be expected through the processes of stellar nucleosynthesis. As we said in the previous section, stellar nucleosynthesis involves combining the nuclei of smaller, simpler elements to form larger, more complicated ones.

The excess of carbon in the universe went unexplained for some time, given that the stellar processes that formed it should not have been able to produce so much. It was discovered that the process of forming carbon relied on a very unexpected coincidence in the resonance of the smaller elements that were combined to create it and the final energy state that the carbon was in after the nucleosynthesis was complete.[4] This highly unexpected resonance led some scientists to think that perhaps there was some guiding force in the universe, given that the likelihood of such a resonance was so small and that this tiny coincidence formed sufficient volumes of carbon to provide enough organic molecules to form living organisms. Carbon-based life would likely not have existed had the resonance been slightly different.

> Coincidental confluences of circumstances can profoundly affect the course of evolution of a complex system.

There is also the fundamental question of whether organic molecules were first synthesized on Earth or arrived on Earth in meteorites. Both explanations may be true. Indeed, many meteorites have struck Earth and have been shown to contain a large number of organic molecules, including some as complicated as amino acids which are the building blocks of proteins. Even more surprising is that meteorites have been found to contain amino acids that do not occur in biological systems on Earth.[5] This suggests that as well as a highly diverse family of chemical elements being generated in space, highly diverse organic molecules are also being generated and find their way to earth.

As for the formation of organic molecules on earth, the Miller–Urey experiments give us an insight into this. In 1952, Stanley Miller and Harold Urey at the University of Chicago performed an experiment where water, methane, ammonia, and hydrogen (all inorganic molecules that would have existed on primordial Earth) were sealed inside a flask and heated to form a vapor. This vapor was then exposed to an electrical spark, in much the same way as water vapor in the earth's atmosphere is sometimes exposed to lightning. Analysis of the compounds formed by this process identified several amino acids, the building blocks of proteins.[6]

From these findings, we can see that organic molecules formed both in space and on the Earth. The next step is to look at how these organic molecules formed early life, a process known as abiogenesis.

There are many competing theories to try and explain how isolated organic molecules could combine so that the resultant product could be called a living organism. Even the simplest lifeform must, in some way, be encapsulated from its environment. It must also be able to metabolize to create energy and, in some fashion, reproduce itself. Stuart Kauffman has shown how certain simple organic molecules can also catalyze reactions to create new versions of themselves. This has been demonstrated in RNA, DNA, and other organic molecules essential to cellular life. Kauffman refers to these as autocatalytic sets.[7]

Certain systems with sufficient complexity can create themselves and build order.

As for encapsulation of these autocatalytic sets of organic molecules, we need to consider the boundary layer of a cell. Biological cells are characterized by a lipid (fatty) membrane much like a sealed bag containing the cell's contents. The lipid nature of this membrane means that it won't dissolve into the surrounding aquatic environment (given that these early cells likely formed in a watery environment). We now have to imagine a scenario where these lipid membranes form as sheets and then encapsulate autocatalytic sets of organic molecules. Kauffman highlights that one of the most appealing theories is presented by Bruce Damer and David Deamer in their Hot Spring Hypothesis for the origin of life.[8]

This story unfolds in freshwater pools on primordial Earth. These pools have already been seeded with inorganic and organic molecules through the processes we described. Damer and Deamer consider the environment at the edge of such pools. Here, freshwater is heated and cooled due to the normal heating of the Earth by the Sun and its subsequent cooling. With the changing temperature, the edge of this pool of water moves forward and backward as the freshwater evaporates and then condenses. It is at this moving edge that the organic and inorganic molecules start to form more complex structures. The process of hydrating and dehydrating the jelly-like conglomerations of molecules at the edges of these pools could lead to the encapsulation of an autocatalytic set of organic molecules by a lipid sheet. We now have an encapsulated autocatalytic set that is distinct from its environment and has the potential to move and carry its autocatalyzing organic molecules into the rest of the pool. As can be seen here, the simple effects of heating and cooling caused by the Earth's normal day/night cycle were the driving force behind an early step in biological evolution.

The interaction that occurs between simple systems can lead to an unexpected increase in the complexity of one or both of them.

It is tempting to go into more detail on this particular theory simply because it is so interesting. However, it is enough to describe it as an important step from static, albeit organic molecules to something which could reasonably be called a living organism.

From there, we enter the realm of Darwinism and evolution by natural selection. Those primitive monocellular organisms with autocatalytic sets best suited to the environment they encountered may have persisted, while those with less advantageous internal chemistries died out. As environmental pressures led to evolutionary selection, these monocellular organisms may have combined and exchanged their primitive version of what we would call genetic material. In so doing, they became more complex.

> To be successful, evolving systems need to develop to match the specific challenges that their environment presents to them.

We know that monocellular life eventually led to multicellular life which then blossomed in diversity as it was first exposed and then selectively adapted to the diverse physical environments on planet Earth. From atoms formed in the hearts of stars, self-organizing complexity has resulted in a biosphere of breathtaking diversity.

2.3.3.1 Minimum gene sets

There is one question that we have not touched on yet. We have talked about a conceivable way that organic molecules can create and replicate themselves (autocatalytic sets) and how these processes could eventually create a monocellular organism, which we might reasonably call "alive." But what is it that a cell needs to survive and replicate?

Research into the minimum length of the genetic material needed for such an organism to be viable led to the concept of a minimum gene set.[9] By selectively adding or deleting genes from the genome of a simple bacteria, researchers could see what combination of genes led to a viable organism, i.e., one that would survive and replicate, and what genes they could delete without affecting this viability.

An early study that adjusted the genome of one of the simplest monocellular organisms found that only 382 out of the 482 genes known to make proteins (the molecules used to carry out all biochemical processes in the body) were essential.[10] Interestingly, a quarter of these essential genes were of unknown function. The others were required to encode proteins to allow the organism to survive by metabolizing food and replicating.

Best estimates suggest that the human genome contains about 20,000 protein-coding genes. When we consider the biological capabilities of our species, we can begin to see that increasing the complexity of our biology requires an increase in the complexity of the coding that allows that biology to be viable, i.e., to survive and replicate. As we will see later, when we want to design viable systems, we need to be aware that essential "genes" give rise to essential "structures" that endow a system with viability.

2.3.4 Entropy and constraint

In the space of a few thousand words, we have gone from the beginnings of the universe to the rich and diverse biosphere that surrounds us on this planet. We could justifiably say that the early steps in the evolution of reality were driven by physics, the intermediate steps were driven by chemistry, and the later steps were driven by biology. The increasing complexity of the universe could be divided into these three epochs; physical, chemical, and biological.

I am skeptical about these divisions for the same reason as I think we need to take a consilient view of the dynamics of the interlocking systems of reality. As the natural sciences evolved, it was understandable that greater specialization and an increasing breadth of knowledge would necessitate splitting up areas of study. There would reach a point where even the most autodidactic polymath wouldn't be able to take in the amount of information being generated in the field of natural sciences, and so would need to focus more on what she might perceive as groups of similarly aimed areas of study. Some natural scientists might focus on plant and animal studies; others might be more inclined to investigate astronomical phenomena, while others might prefer to experiment with beakers of chemicals in a laboratory. All would have been called natural scientists at some point in the past, but as the field grew, so branched off the biologists, physicists, and chemists.

What if we ignored these different "flavors" of natural science? Instead of seeing processes through the lens of physics, chemistry, or biology, what if we just viewed whatever object or reaction between objects as a question of systems? Instead of atoms, molecules or organisms existing and interacting, we simply have components existing and interacting as systems.

Without referring to any physical, chemical, or biological specifics, let us go from the Big Bang to the present day by imagining the universe's evolution as the evolution of a complex system. We've already tried describing this evolution in terms that are inextricably grounded in their own specific part of the natural sciences. Still, we may be able to relate the universe's evolution to something more general.

Entropy is often defined as a measure of disorder in a system, although disorder is a slightly misleading term. Imagine you sprinkle sand from a height onto the floor. When you look down at the sand, it will appear that you have covered the floor reasonably evenly. However, when you get closer, you will see that the particles of sand are not neatly arranged into an ordered pattern. They are randomly distributed but in a relatively even way.

A better way of thinking about entropy is to think about it as a measure of how compartmentalized the energy in a system is. More specifically, entropy is about how different the level of the compartmentalized energy is in one block of matter (solid, liquid, or gas) compared with the other blocks of matter it is in contact with. For what follows, you can think of the energy of a compartmentalized block of matter as its temperature. This will make a bit more sense shortly.

The second law of thermodynamics tells us that entropy will increase in a closed system (i.e., a system without any input of energy from outside). A classic way of imagining this is to think of a jar of hot water into which you place a few ice cubes. We know intuitively that the ice cubes will melt, and then instead of a hot region (the hot water) and a cold region (the ice cubes) within the jar, the ice cubes will melt into cool liquid water, which will then absorb some of the heat from the hot water. The process of melting the ice cubes will cause the liquid water already in the jar to cool down slightly, and finally, all the water will be at the same temperature. The energy difference between the coldness compartmentalized in the ice cubes and the hotness compartmentalized in the water progressively decreases. Instead of having two organized, discrete (or "ordered") bodies of matter (the ice and the hot water) at two different temperatures, without our input, all the molecules from the ice and the water merge together, sharing their energy equally. All of that previously compartmentalized energy in two different regions is now distributed evenly throughout the jar. The ordered boundary between ice and water has disappeared, and the heat energy is evenly spread throughout the water. Remember, just because the energy distribution is even does not mean that the energy distribution is ordered. We have gone from having two ordered regions with different energy levels respective of each other to a disordered distribution, albeit an evenly disordered one. This is entropy.

Should the ice cubes spontaneously reform within the jar, we would rightly be amazed. It is also an impossibility according to the second law of thermodynamics. To separate the water in the jar into two distinct temperature zones, we would need to put energy into the system.

We now come to the difference between local and total entropy. I can take some of the now tepid water from the jar and put it into an ice cube tray in my freezer. Using the energy of the freezer, I am now cooling the water so that when it is compared to the water that remains in the jar, we can see that we have again got two blocks of compartmentalized matter within which the energy levels are different. We could then repeat the process of dropping the ice cubes into the jar of tepid water.

How does the process of taking a portion of the tepid water and cooling it down impact our understanding of entropy? Surely we have now decreased the entropy in this particular system by recreating the difference in temperature that existed at the start. At a local level, that is correct, but there is one important feature of the process that allowed us to push back against the tide of increasing entropy: we used the freezer's energy to do it.

If the freezer had no energy, over time, much as the ice cubes in hot water did, the temperature difference between the freezer and its environment would reduce to nothing. It is only when the freezer is powered that we have the potential to change the compartmentalization of energy of a unit of matter with respect to its surroundings. It should be noted that this idea works just as well if we replace the concept of cooling a block of matter with the concept of heating a unit of matter, i.e., we could replace the freezer with a kettle, and the principle still stands. In both examples, we are using electrical energy to raise or lower the temperature of a block of matter with respect to its surroundings (local entropy decreases as we have now compartmentalized a specific energy level within the block of matter with respect to its surroundings). Once we remove the electrical energy from the freezer or the kettle, the water inside will gradually cool or warm until it is the same as its surroundings; the compartmentalized, ordered energy difference between the water and its surroundings cannot be maintained without external help, and so this energy becomes randomly spread out between the water and its surrounding environment until the temperature difference is zero. This is entropy at a local level. We can use energy from outside the system to reverse entropy. Still, as soon as that energy is cut off, the spontaneous process of randomly distributing all the energy compartmentalized at different levels in different parts of the system begins again in earnest.

Suppose we zoom out from the local level. In that case, we need to consider where the electricity came from that allowed us to reverse entropy at a local level by either freezing or boiling water so that it then had a different energy level to its surroundings. The process of creating electricity (i.e., a force which is characterized by its high energy) requires work to be done. Work is any sort of chemical or physical process. Coal or gas is burned, creating heat that is converted into work in generators, which then increases the conducting system's energy levels (which also releases some heat). Again, it seems like we have decreased entropy by compartmentalizing energy. If we zoom out again, we need to consider the fuels we burned. Coal and gas possess energy because of sunshine. The energy radiated toward the Earth from the Sun allowed for the processes that enabled the trees to grow and compartmentalize the energy that we would later use as fossil fuels. Again, we seem to have locally defied entropy.

Finally, we get to the Sun itself. The Sun is doing work all the time as it fuses hydrogen nuclei to form helium and energy through nuclear fusion. This work can only be achieved because it has a lot of fuel (hydrogen) and because it has a powerful enough driver of fusion (its own gravity). Eventually, the Sun will use up all its fuel, and fusion will stop. This is broadly true for all stars in the universe, although the manner of their demise depends on other factors such as size and composition. These stellar remnants will continue to radiate heat until they become as cold as deep space. The chances are that if it still exists, the human race will be quite a different species from what we are now and unlikely to require coal to keep the lights on (wherever we are in the universe). The crucial point is that now that we no longer have the primary energy source that allowed the trees to grow and so locally decrease entropy by building order (the structure of the trees). Once we have taken advantage of all the compartmentalized, ordered stores of potential energy, be they fossil fuels or radioactive fuels or winds driven by solar warming of the earth, entropy will inevitably increase. Gradually all the energy in existence will become randomly and evenly distributed throughout the entire fabric of reality. Without a way of compartmentalizing energy and then using it, no physical, chemical, or biological work can be done, and we arrive at the so-called heat death of the universe. Entropy will win in the end.

> We can decrease entropy locally by using energy from outside our system. Still, entropy in the universe will always increase toward the final heat death of the universe when all energy is evenly spread throughout reality.

You might ask yourself why we are spending so much time looking at entropy. It was necessary to go into some detail because it is a subject that is not widely understood. Still, it does impact how we understand the unfolding of systems within the universe. We said we would temporarily dispense with the distinctions between physical, chemical, and biological systems, and so to talk about systems in a general way so that we can get some insight into their similarities, no matter what domain there are in, we need to talk about these universal principles.

As we looked at entropy, an idea that came up frequently was compartmentalization. By decreasing entropy at a local level and compartmentalizing different blocks of matter with varying energy levels, we can later rely on exploiting the energy that flows between these two units of matter as entropy occurs in order to do work.

When we have compartmentalized energy, we then want to use it to perform a task. In the case of the ice cube in hot water, the compartmentalized energy is released in every direction, and the water cools slightly. This may have been what we wanted, but the hot water would have cooled itself even without the ice, so all we have done is accelerate the process. Suppose we drop the ice cubes into water already at room temperature. In this case, we are releasing this compartmentalized energy difference to lower the temperature of the water with respect to the environment. We will temporarily decrease the local entropy of the water in the glass with respect to the environment, and we can enjoy our nice cool drink. Would the effect have been as good if we had just placed the same number of ice cubes outside the glass? The answer is no.

Exploiting compartmentalized energy requires us to *constrain* the energy release in some way. Ice sitting in an ice bucket on the bar may well cool your drink by an infinitesimal amount simply by virtue of the fact that the heat in the room decreases ever so slightly as some of the compartmentalized energy within the warm air seeps into the ice bucket to raise the temperature of the ice. The temperature of the ice and the temperature of the room are destined to equilibrate, and if your drink is at room temperature, that too will decrease by the same tiny amount as the room temperature. However, we could put that ice to better use.

Suppose we place the ice into our drink instead. In that case, we will achieve a much greater benefit because we are constraining these ice cubes with their compartmentalized energy states that are different from their surroundings. Entropy wants the higher temperature in the room to melt those ice cubes so that all the energy gets evenly spread throughout the local environment. However, by putting the liquid in our drink into close proximity with the ice cubes and by having the whole thing in a glass (or even one of those clever insulated glasses), we can constrain the process at an even more local level so that instead of the ice causing a tiny drop in temperature of a large volume of air, it is causing a large drop in temperature of a smaller volume of liquid.

> Constraining compartments within systems mean we can more efficiently exploit compartmentalized differences in energy levels between blocks of matter to create order (i.e., we can locally decrease entropy).

The second law of thermodynamics also explains why we can never create a perpetual motion machine. Work processes can never use energy in a way that is 100% efficient. In a simple example, friction in a wheel that we have spun by hand will convert some of the energy we imparted to the wheel into heat, and this will be radiated away. Even the energy lost by processes that might not necessarily have heat as an immediate byproduct of their intrinsic, possibly tiny, inefficiencies will ultimately become heat, which radiates away from the local environment and contributes to a general increase in the overall entropy of the universe.

2.3.4.1 Emergence

Entropy and constraints will become important, but before we go on, we need to address the tricky subject of emergence, a term that some of you might be familiar with, particularly those who have had some exposure to complexity science previously.

When complexity science began to be viewed as a discipline in its own right, emergence became a buzzword within the field. Broadly speaking, emergence is the "phenomenon" where groups of separate but similar entities come together and start exhibiting ordered behavior in a way that does not seem predictable given the properties of the individual, i.e., the potential for that behavior doesn't appear to be encoded into the individual.

One of the classic examples used to describe the phenomenon of emergence is that of slime molds. Slime molds are a diverse group of monocellular organisms that exhibit interesting behavior. For example, slime molds can aggregate into a larger multicellular organism that resembles a slug. This behavior can be triggered by a lack of food in the surrounding area. Rather than individual monocellular organisms trying to translocate themselves to an area with a richer food supply, it is more efficient for the monocellular organisms to aggregate into a larger multicellular entity which can then translocate itself before redistributing itself again as monocellular organisms or growing upward into fruiting bodies to distribute spores as part of its reproduction cycle.[11]

Research into other aspects of slime mold behavior has also yielded interesting observations. For example, when researchers placed small morsels of food on a Petri dish in an arrangement that resembled the larger metropolitan districts that are part of the city of Tokyo, it was discovered that the slime mold distributed itself throughout the Petri dish in a way that replicated the layout of the Tokyo Metro system.[12] Explaining how monocellular life forms can exhibit this complex behavior has been attributed to the phenomenon of emergence.

At first glance, it almost seems impossible. How could such a simple organism ultimately distribute itself in a pattern similar to the metro system of a large metropolitan city? Something must be driving the simple toward the complex. This idea that large conglomerations of individual components, especially biological ones, can exhibit behavior that doesn't appear to be predictable from what we know about the individual components and that such behavior is somehow "emergent" comes up repeatedly in many areas of complexity science. The emergence of culture, financial markets, language, and religion, to name a few phenomena, seems to be entirely unpredictable given the nature of the individual units that form the communities in which these phenomena are seen, in these cases, humans. Outside of the human world, termites are seen to produce strange-looking mounds (sometimes called cathedrals due, one imagines, to their Gaudi-esque features) that ultimately help cool the termite colony. This semiarchitectural creation is sometimes called emergent because where in an individual termite's genome is the capacity to build such a thing?

In neurobiology, the phenomenon of consciousness is labeled as an emergent phenomenon as it cannot be predicted (or "pre-stated" to use another popular term in emergence circles) based on our knowledge of individual neurons which are, in some way, the "individuals" that go to make up the nervous system.

Because of all these weird and wonderful observations, emergence has somehow come to be seen as a phenomenon in itself.

It may be that emergence is actually not a useful way of describing the behavior of complex systems. It may be that we have created this imaginary phenomenon of emergence to account for the fact that we do not fully understand the current workings or the potential evolution of the complex systems that we are observing. Take the example of the slime molds. I cannot deny that it must have been remarkable for the scientists studying slime molds to see the monocellular organisms aggregate together into a larger, slug-like multicellular organism to move and find more food. How on earth could these monocellular organisms have managed to coordinate themselves in such a way?

My answer to that is quite simple. They can because they did.

We might not completely understand every step of the process by which they aggregated into a larger multicellular organism, but we can make a decent guess at the initial conditions. Lack of food leads to biological stress. Biological stress leads to the release of chemical signals. Chemical signals trigger biological changes in the cells. These changes make it amenable for cells to combine. Chemical signaling flowing between these aggregated cells becomes more efficient than it would be through a widely distributed cellular network. In this way, the larger multicellular organism can move in a coordinated fashion.

Suppose my interpretation of emergence, or the lack thereof, is correct. In that case, we could almost say that emergence is being used in the same way to explain the unexplainable in the same way as mythology was used to explain the unexplainable in early human civilization. Did the first humans seeing an eclipse feel any less wonder than the observer seeing slime molds aggregate into a slug to move and find food?

We were ultimately able to explain why eclipses occur, and, in the same way, I have little doubt that we will be able to explain many of the so-called emergent properties of complex systems.

The other feature of emergence that troubles me is this idea of predictability in trying to label phenomena as emergent. Scientists often say that the outcome was entirely unpredictable based on the initial state of the particular community members that gave rise to the "emergent" phenomenon in question. Given what we generally know about systems so far, we know that most complex systems are open systems that exchange energy, matter, and information with their surroundings. If this is true, why should we be amazed when we cannot predict the future? A complex system that comes together and evolves is constantly exchanging matter, energy, and information with its surroundings. Perturbations in the surroundings alone are enough to make a complex system evolve in a way that we might not expect.

In the same way, the ultimate building blocks of any complex system, the atoms and subatomic particles that form the structure of matter are also sources of uncertainty and possibility due to quantum mechanics. Why then should we be surprised that we cannot predict the future of a complex evolving system on this basis alone? If a system were closed, as in the examples we used when we talked about entropy, I would be amazed if any sort of complexity formed in the absence of energy coming in from the outside as doing so would violate the second law of thermodynamics. However, in an open system, we can see that the so-called emergent phenomena are simply ways that complex systems constrain and use energy to form order and, in this way, locally decrease entropy.

In looking at complex systems, I have never found emergence to be useful in understanding the deeper functioning of the system in question. It seems an arbitrary placeholder word that we use when we might not entirely understand how a complex system is evolving. It may be very interesting that a slime mold can transform itself into a representation of the Tokyo Metro system. Still, I am guessing that the designers of the Tokyo Metro system weren't using slime molds to help them decide where to lay the tracks. The designers used their own analysis and their own faculties to make those decisions for themselves. Because we know more about those analysis methods, and indeed, the faculties and capabilities of the humans involved, we don't call the laying of the Tokyo Metro system an emergent phenomenon. This would seem to be an excellent example of how our silo thinking goes contrary to the ideas of consilience in that we have two systems, both operating on very different scales, which have evolved in various ways to give the same overall result. The fact that the layout calculated by the designers of the Tokyo Metro system and the layout that developed in the slime mold were the same should suggest to us that perhaps there is a more profound unified process going on than we might initially think. This is at the heart of consilience.

2.3.5 A consilient approach to the evolution of complex systems

Before we embark on looking at the complex systems that are probably the most interesting to us, i.e., the ones involving humans, it is worth considering how the concepts of entropy, order, constraints, and information relate to what we know about the evolution of the universe so far. Again, we will disregard the arbitrary boundaries meant to distinguish physical, chemical, and biological processes. Instead, we will look at this evolution purely in terms of systems.

1. The universe came into existence with a quantity of energy (some of this energy in the form of matter according to the relationship between mass and energy as described by Einstein).
2. The formation of matter constrained energy within itself given that matter can be transformed into energy, such as when nuclei combine during nuclear fusion and some of their collective mass is converted to energy.
3. Energy needed to be applied to initiate the reactions that combined simpler components into more complex systems. This energy may have been potential energy already existing in the simpler components.

4. Some reactions may release more energy than they would seem to consume, but this is due to a mass/energy conversion or the release of potential energy contained within the components.
5. Despite the total entropy of the universe increasing, the process of compartmentalizing energy within increasingly more complex systems means that it is possible to decrease local entropy.
6. The encoding of information is also an energy-driven process and can be seen as a local reduction of entropy.
7. When systems reach a certain level of complexity, energy is required to maintain the boundaries of the compartments that have a different energy state to their surroundings.
8. Energy is needed to perform work.
9. In a system, compartmentalized energy is best used by constraining the path of the energy released so that it flows through and energizes the work process that is required.
10. Work processes are never 100% efficient which means there will always be an energy loss, the heat of which will ultimately contribute to the increasing entropy of the universe as a whole.

As we will next be looking at how human beings arrange themselves into increasingly complex systems, let us just put the above list into the context of planet Earth. The existence of every piece of matter, energy, or information on earth, no matter the magnitude or how simple or complex the system in which it finds itself is a result of energy-driven work, which decreases the local entropy and so builds order. Energy radiated in from outside the locality (the planet), whether used directly or indirectly from ancient stores of this energy, allows this ordering process to continue and allows the boundaries of the compartments to be maintained. It is like we are surfing on the wave of the Big Bang, picking up energy to allow us to pick up speed and perform increasingly more complex maneuvers. Although we are locally picking up energy to use, the wave as a whole is dissipating its energy and will eventually fade into nothingness.

2.3.6 Societal systems

And so, we find ourselves as humans surfing down the face of the wave generated by the Big Bang. Luckily for us, there is an abundance of energy close at hand which we can use to build the ever-increasingly complex systems which constitute our society on Earth.

Consider the arrangement of the human beings on the face of the planet. Historically speaking, the main boundary to the various compartments containing human beings was geographical. Valleys, rivers, or even coasts all acted as boundaries that constrained human beings and made it less likely they would move from their current location to a different one. The majority of the most obvious, persisting perceived differences between humans stem from this geographical compartmentalization. However, human beings went further and employed energy to form ever smaller compartments for themselves within the larger compartments provided by nature. A village constructed on a piece of land would serve as a way of separating those human beings who lived within the village from those who lived outside it.

Observation and knowledge acquisition from previous generations then engendered beliefs about the nature of the world and the place of humans within it. Different observations and the passing down of different knowledge may have led to a relatively homogeneous group of geographically constrained humans to start believing in different things. So the group separated and compartmentalized with respect to religious faith or lack thereof.

Given that we framed the universe's evolution in terms of energy, information, matter, and entropy, an interesting question to consider is whether cultural, religious, and social systems operate in the same way. What would happen over a period of time if energy was not expended in generating and transmitting information to individual human beings. Maintaining a social or cultural group circumscribed by a set of inherited beliefs requires the continual application of energy in the same way as maintaining the boundaries of a compartment of matter with a higher or lower temperature requires energy to prevent entropy from increasing. Suppose one generation of cultural or religious leaders decided not to expend energy in engendering beliefs in the generation that followed. The boundaries that circumscribe that particular group might start to break down. Through a process akin to that of entropy, will there come a time when human beings cease to recognize cultural, social, and religious differences between them? It may be that the historical, energy-driven formation of boundaries that compartmentalize human beings into separate cultural, social, and religious groups will eventually dissipate in much the same way as ice cubes melting into water.

Let's now focus more on the formation of organizations. The ability of human beings to come together to form systems is often based on similarity. For example, for a system to function well, the humans involved need to have a commonly understood way of transferring matter, information, and energy between them. Acquiring the knowledge of how to do this is another energy-dependent process.

More specifically, let's consider a group of humans who come together to construct a mechanical watch. Before beginning the actual construction, the humans must come together and exchange information to allow them to plan out the tasks that are to follow. As knowledge is shared between them, the overall complexity of the system they constitute increases as this information is encoded in their brains. It may be necessary for one human who has particular expertise in one area of watchmaking to impart this information to someone else. Again, energy is expended to increase complexity as the second human learns something new.

In summary, even before the humans access the tools and materials needed to build this watch, they have formed a social system of increasing complexity just at the planning stages.

2.3.7 Informational systems

The evolutionary development of disparate humans into what could be called societies and the subsequent development of those societies suggests that there may be a less concrete class of system embedded within societies. In the previous example, we described how humans coming together resulted in the exchange or generation of knowledge. It also results in the *storage* of knowledge. The recording, transmission, and accessing of knowledge are energy-dependent processes, and energy must be expended to maintain the compartments within which information is stored. Books will eventually crumble to dust. Hard drives will eventually fail. Information that is transmitted multiple times is susceptible to degradation unless it is refreshed through some process that, in itself, will be energy-dependent.

Understanding the relevance of informational systems will be important further on in the book, but at this stage, it is enough to define what we are talking about; it is not the physical nature of the storage medium but rather the organization of the information itself.

2.3.8 Technical systems

So far, we have focused on systems that involve biological, organic entities. We cannot have a complete picture of the various classes of systems without considering purely technical systems. These are systems constructed by human beings but do not rely on human beings after their initiation. A mechanical watch is an excellent example of this, and some of the internal components of such a watch are shown in Fig. 2.3.2:

Once the watch has been built, the human initiates it by setting the current time and winding it up. The process of winding compartmentalizes energy within the mainspring. The mainspring is connected to the gears that drive the hands around the face of the clock. If it were that simple, as soon as you had finished winding the watch, all the energy in the mainspring would be released, and the hands would whizz around the face of the watch until all the energy compartmentalized in the mainspring had been expended. Something is needed to *constrain* the release of energy for the watch to perform its function, namely keeping accurate time. The escapement mechanism of the watch constrains the release of energy from the mainspring by locking the gears that would ordinarily drive the hands around the face of the watch. If the escapement remained static, no energy could be released from the mainspring. To unlock it in a highly precise way, the escapement is connected to a balance wheel, which partially rotates in one direction before swinging back in the

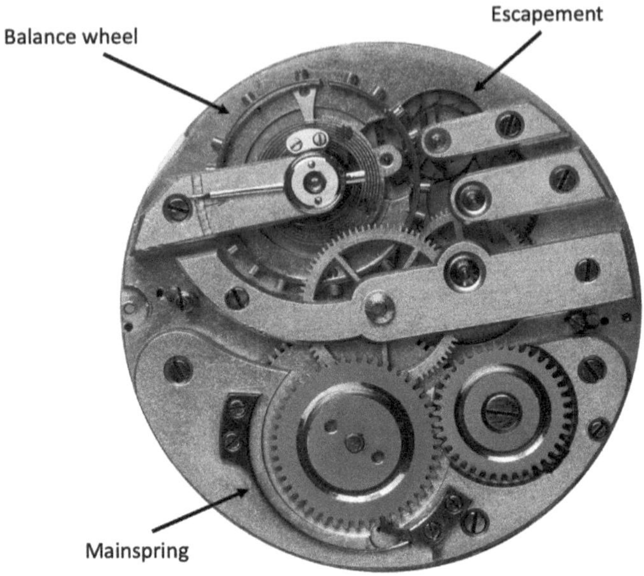

FIGURE 2.3.2 Schematic of a mechanical watch. *Image copyright Tobias Arhelger (www.shutterstock.com).*

other direction thanks to a small spring. The timing of this swinging backward and forward gives the watch its timekeeping ability.

The movement of the balance wheel unlocks the escapement, which then permits the gears to move as a tiny amount of energy is released from the mainspring. Although most of the energy is used to drive the watch's hands, a tiny amount of that energy passes back through the escapement and into the balance wheel to keep it swinging regularly and permit accurate timekeeping. Without this tiny nudge to keep the energy up in the balance wheel, the natural loss of energy through friction and heat would soon stop the movement of the balance wheel, and the whole mechanism would cease to function.

This is an excellent example of how the constrained release of energy allows a system to perform work. After its initial winding and setting, the mechanical watch will operate in a manner that does not require human input until the point where the mainspring is exhausted, and the human must wind the watch up again to maintain its operation.

An interesting feature of mechanical systems is that we cannot really call them complex. Given that our only inputs are the initial setting of the time and the subsequent winding of the mainspring, the technical system in question will only do what it has been designed to do. It will not unexpectedly start talking or display the latest sports results. All it will do is function (to a variable degree of accuracy) until the point where it stops functioning. Based on the Cynefin framework, we could say that a technical system can never be more than a complicated one. The functioning will always be understandable and almost entirely predictable to someone with sufficient knowledge of the technical layout of the system.

2.3.8.1 Theoretic minima

If we consider a mechanical system designed to perform a function, we can begin to think about how we might most simply go about the process of constructing such a system. We can consider the "theoretic minimum" of the system. As I see this concept, I think it is only currently applicable to technical systems, i.e., ones that are entirely engineered and are designed to fulfill a specific function.

Let us say that a watchmaker has been given a commission to make the simplest watch possible for a specific client. The client may not have specified any requirements about durability or repairability nor any specific materials they would like used. The watchmaker now must consider the design of the watch. She may decide to impose some arbitrary conditions to protect her own reputation. For example, she may want to ensure that the watch has an expected lifespan of 50 years and can easily be repaired during this time. This requirement will start to specify the types of material that she can and can't use in the watch's construction.

She will also have to consider the function that the watch is designed to perform. The time will need to be readable and, depending on the client's visual acuity, this will determine the size of the watch face and the hands. From there, the movement can be designed to provide enough motive energy to move the hands. She could minimize the materials required to build the watch by including a tiny mainspring, but this would mean the client would need to wind the watch up every hour. Perhaps she thinks it best to put in a mainspring big enough to store enough power to keep the watch working for a day before it needs to be wound up again.

Now let's suppose that her next client is incredibly picky. He knows exactly every parameter he wants to be designed into this mechanical watch, from readability in low light to accuracy within 0.5 seconds per day and a life span exceeding 100 years. The watch needs to have a big enough mainspring so that it only needs to be wound every 3 days, and it also needs to have a calendar. Whenever the watchmaker asks him if he prefers this feature or that feature, he always has a demand.

She now has this detailed, interminably long list of parameters that this technical system needs to fulfill. The concept of a theoretic minimum is that there exists a combination of materials and design, which is the absolute minimum mass of components in the watch and requires the absolute minimum amount of energy to run. It might not be possible to know what it is, but it theoretically exists.

For a mechanical system like this, it would be sensible to say that the lowest mass of material would also likely require the least energy. You might have to consider two theoretic minima in different mechanical systems—a theoretic mass minimum and a theoretic energy minimum. We will return to this idea of trying to achieve theoretic minimum in a system later in the book as ideas similar to this sometimes crop up in other system types and can cause severe problems.

Theoretic minima are only useful in technical systems with clearly defined functions.

2.3.9 Sociotechnical systems

To complete our journey through the different classes of systems, we must finally consider systems that combine human, technical, and informational components. These are known as sociotechnical systems, and they have their own chapter later in the book because they are so essential to all manner of systems designed to perform some sort of productive work. However, just briefly, a sociotechnical system is a combination of human components, technical components, and informational components (laws/rules/procedures/software) which come together to accomplish a common purpose. In this chapter, we have looked at certain features, dynamics, and concepts associated with social systems, biological systems, physical systems, and technical systems separately. When it comes to sociotechnical systems, we now need to start combining this information to help us understand how they work and give us some insights into how we can best manage them. However, there is one very important subclass that we glossed over quite quickly in our rapid tour from the birth of creation to the watchmaker's bench. It will be essential to our understanding of sociotechnical systems later and will give us insights into how we can best manage complexity. This subclass is the human nervous system.

Chapter 2.4

Neurobiological systems

(The brain is) a machine for jumping to conclusions.

Daniel Kahneman, Nobel Prize winning psychologist

Chapter outline

Designing and Managing Complex Systems. https://doi.org/10.1016/B978-0-323-91609-7.00005-6

2.4.1 Introduction

You might be asking why this particular subclass of the broader class of biological systems merits a chapter of its own. We barely mentioned it as we looked at the evolution of the universe up until the present day, but that is only because it is so essential to the later parts of this book.

The other reason it merits a chapter of its own is that the human nervous system is the crowning glory of evolution. That may seem like a sweeping statement, but it truly is. Nature has given us an awesomely complex system of our very own that we can use to help us deal with all the novel, diverse, and dynamic environments that we find ourselves in. If we can gain a deeper understanding of how our nervous system actually accomplishes this, we can see how its processes can be applied in different ways to let us design and manage complex systems more successfully.

2.4.2 Structure and function of the nervous system

The human nervous system has three interconnected subsystems

1. The central nervous system (CNS) – Comprising the brain and spinal cord.
2. The autonomic nervous system (ANS) – Comprising certain brain areas (particularly in the area where the spinal cord joins the brain) along with two series of linked "ganglia" (clusters of neurons) that run either side of the spinal cord. The ANS, much as the name suggests, deals with the body's automatic functions such as heartbeat, breathing, digestion, etc. It is also involved in the fight-or-flight response. The ANS itself has two subdivisions:
   a. Sympathetic – Important in regulating heart rate, breathing, and body temperature (so-called "homeostatic" control—more on that later). Becomes more active in fight-or-flight scenarios, and activation of the sympathetic division generally leads to an upregulating of bodily activity.
   b. Parasympathetic – Works in an almost contrary way to the sympathetic subdivision and tends to promote a "breed and feed" response associated with recuperation, digestion, sexual arousal, and the body's voiding mechanisms.
3. The peripheral nervous systems (PNS) – All of the wiring that connects the central nervous system to the sensory and motor organs of the body.

> The nervous system is divided into functionally integrated sections that allow it to perform its functions.

We will look at the autonomic nervous system in more detail later in this chapter. However, for now, let us look in more detail at the general structure and function of the rest of the nervous system, with particular emphasis on the central nervous system.

2.4.2.1 The central nervous systems

The human brain is composed of brain cells. These cells are called neurons, and there are approximately 86 billion of them in the human brain. Before we look in more detail at how these neurons work, let's consider the brain as a whole. The human brain is broadly divided into functional areas. Fig. 2.4.1 shows the brain as seen from the left side, and Fig. 2.4.2 shows the midline structures of the brain:

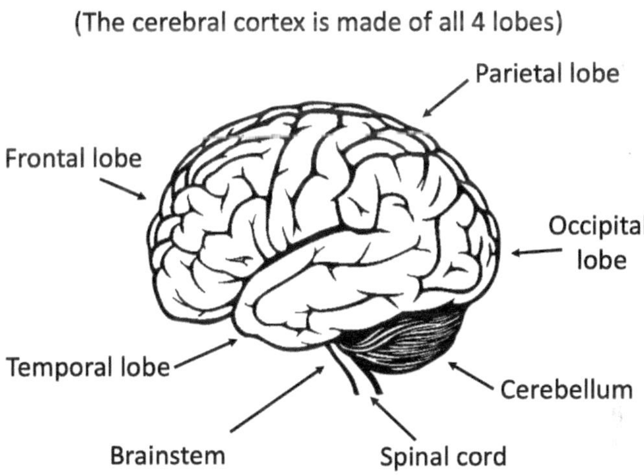

FIGURE 2.4.1 Brain anatomy as seen from the left side. *Image copyright takito (www.shutterstock.com).*

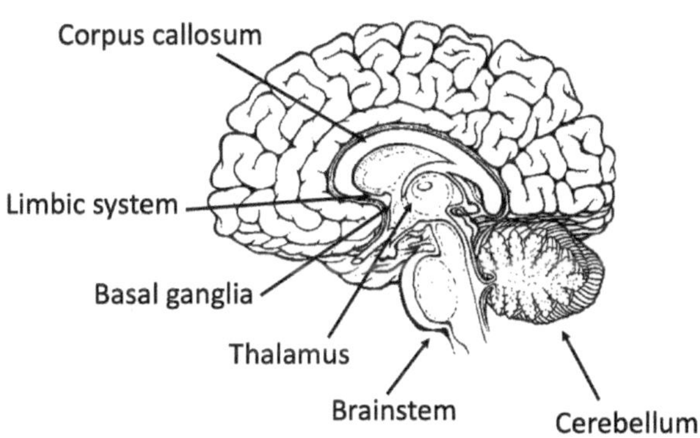

FIGURE 2.4.2 Midline brain anatomy. *Image copyright Denis Barbulat (www.shutterstock.com).*

The upper part of the brain, the cerebral cortex, is highly developed in most mammals but may be considerably smaller or even absent in other types of animal life. In most mammalian brains, the cerebral cortex is divided into two halves, known as hemispheres. The brain's left and right parts are linked and communicate through a bridge of nerve tissue called the corpus callosum.

Most animals with a nervous system have part of the brain dedicated to regulating basic physical functions, for example, heartbeat, respiration rate, sleep/wake cycles, etc. Higher mammals such as human beings went on to develop the more complex cerebral cortex to achieve more advanced functions, for example, more advanced forms of communication, imagination, and creativity, the ability to plan and execute complex behaviors (known as executive function), and the ability to carry out all manner of social interactions. Explanations for the major areas shown in Figs. 2.4.1 and 2.4.2 are given below:

1. The brainstem, located toward the rear of the brain and leading out into the spinal cord, is the part that we share with most forms of animal life. It is this part of the brain that regulates our most basic physical functions, such as heart rhythm and breathing. Damage to this area is often fatal as it disrupts these essential functions. It is this area that connects to the autonomic nervous system.
2. The thalamus, located at the top of the brainstem and surrounded by the cerebral cortex, acts as the junction box for the brain. Because the brain is divided into several different functional areas, the various inputs to these areas must be routed through the thalamus and their outputs integrated as part of the process of responding to stimuli.
3. The basal ganglia, a collection of brain structures located toward the center of the brain, is thought to act as the central decision-maker, i.e., will choose responses from several possible options. It also has a role in learning procedural motor skills.
4. The cerebellum, located toward the rear of the brain but separate from both the cerebral cortex and the brainstem, is more highly developed in mammals that possess fine motor skills. This part of the brain is involved in adjusting our motor outputs (for example, our hand and finger movements) to achieve a high level of dexterity.
5. The cerebral cortex is more highly developed in humans than in other mammals. It is the largest part of the brain and helps us achieve all the functions that distinguish human beings from other animals. Other mammals possess a cerebral cortex, and its size and organization dictate the animal's intellectual capabilities. It is divided into four main regions, known as lobes:
   a. The occipital lobe, located at the rear of the brain, mainly processes visual information. As the vast majority of the sensory information we

receive is visual, a correspondingly large area of the brain is devoted to processing it.

  b. The temporal lobe, located on each side of the brain (remember, the brain is divided into two hemispheres, and so there is a temporal lobe on both the left and the right sides), is involved in processing auditory information, particularly speech. It also has a crucial role in the formation and storage of memories.
  c. The parietal lobe, located at the top of the brain just forward of the occipital lobe, processes sensory information coming from the body and combines this with information from other brain regions. For example, it would combine the sensation of a cold object in the palm of your hand with the visual information derived from the occipital lobe to conclude that the object is probably an ice cube.
  d. As the name suggests, the frontal lobe is located at the front of the brain and has a variety of functions. A section of the frontal lobe sends instructions to various muscles in the body to allow them to move in a coordinated manner. The frontal lobe is also responsible for executive function, which is to do with planning complex behavior as well as interpreting facial expressions and emotions in other people. It also has a role in moderating human behavior based on socially acceptable norms. Damage to this area can result in loss of inhibitions and a failure to interpret the emotions of others.

6. The spinal cord – this is the main route for information coming into the brain and instructions flowing out. It connects the brain with the rest of the body and can be seen as the main communication network that allows information to be gathered and instructions to be distributed.

7. The cranial nerves – although most of the information comes into the brain through the spinal cord, 12 nerves feed directly into the brain, such as the optic nerves (carry visual information) and the vestibulocochlear nerves (carry both auditory and balance information).

8. The limbic system – this, again, is one of the brain areas that we share with many other animals. It is responsible for processing emotion and is closely associated with the thalamus and the cerebral cortex.

> The central nervous system comprises functionally distinct regions that integrate their activity to fulfill the overall function of the system.

Now that we have covered the basic function and location of various brain regions, we now need to go a step further and attempt to understand how these regions actually *achieve* their functions. To do that, we need to understand the basic building block of the human brain—the neuron.

2.4.2.2 Neurons in the central nervous system

Think of the neuron like a wire. Information is passed from one end to the other. Sometimes the distances can be tiny, a fraction of a millimeter, and sometimes the distances can be large, perhaps up to a meter. Despite their small size, if all the neurons in your brain were laid end to end, they would reach 180,000 km, enough to circle the earth four times. Unlike most wires that we are familiar with in our homes, a single neuron may have up to 10,000 links with other neurons. These links are called synapses, and although the adult brain has about 86 billion neurons, the huge number of links between neurons means there are about 500 trillion synapses. Fig. 2.4.3 shows a diagrammatic representation of a network of neurons and an enlarged diagram of a synapse showing the release of neurotransmitters.

When people are told that the complexity of the human brain exceeds even the most complex computer ever created, they often find it difficult to understand why humans aren't capable of performing billions of calculations per second like a computer can. To understand the reason for this, we need to look at how information is passed along these different types of "wires." Although we learned at school that the brain is an electrical organ, this isn't entirely correct. If it were correct, we would be able to exceed the capabilities of even the most advanced computer because our brains are wired up far more intricately than even the most modern supercomputers. The difference between the wiring in a computer and the neurons in our brain is that information transmitted through computer wires is *electrical* in nature (traveling at between 66% and 97% of the speed of light depending on the nature of the wire), and the information transmitted through neurons is *electrochemical.* This

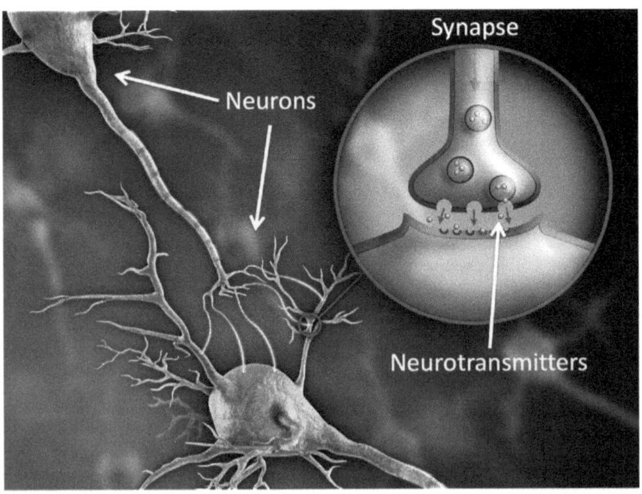

FIGURE 2.4.3 Neurons, a synapse, and neurotransmitter release. *Image copyright Andrea Danti (www.shutterstock.com).*

difference is why the human brain is good at some things and bad others, whereas computers are good at something and bad at others.

Electrochemical transmission in a neuron involves positive and negatively charged chemicals moving in and out of the surface of the neuron. As this happens in one section of the neuron, it triggers the movement of charged chemicals in the next section. In this way, if a neuron is activated, the impulse will travel from the point of activation throughout the neuron and to all the points where it might connect with other neurons. This activation of the neuron is known as "firing." When a neuron fires, the impulse will reach a synapse (the area where it connects with other neurons), and the activated neuron will release chemicals known as neurotransmitters into the tiny space between the two neurons. These neurotransmitters will then have an effect on the second neuron and may cause it to activate as well. In Fig. 2.4.3, the neurotransmitters in the synapse are shown as arrows. The speed of electrochemical transmission is a lot slower than that of electrical transmission. If we consider that electrical transmission in a normal copper wire occurs at about 97% the speed of light (about 291,000,000 meters per second), the equivalent speed in an average neuron is only 100 meters per second. That's only 0.000034% of the speed in an electrical wire. The ability of the brain to do all the amazing things it does is by virtue of the sheer *number* of connections between neurons (i.e., the number of synapses, about 500 trillion). The brain has roughly 500,000 times more connections than the most advanced microchip available.

So, let's summarize this contrast between complexity and speed with the following two facts:

1. Because each neuron may have up to 10,000 connections with other neurons, far more interconnectedness than even the most advanced microchips, the complexity of the human brain's wiring is about 500,000 times greater than that of the most advanced microchip.
2. However, the speed of transmission in the human brain is only 0.000034% of that within a computer.

So, while the human brain can achieve far more complex tasks than a computer, a computer can perform many more simple tasks within a short space of time than the human brain.

> It is high levels of connectedness that can create complexity rather than high connection speeds.

2.4.2.2.1 Types of neurons

There are several different types of neurons that allow the brain to carry out its functions. Sensory neurons bring information into the brain, and motor neurons carry instructions from the brain to all the muscles of the body. Surprisingly, only a very small percentage of the 86 billion neurons in the human brain are sensory or motor neurons. Between 95% and 99% of the neurons in the human brain are of the type known as interneurons. As the name suggests, these neurons form the links between other neurons and are the basis for the highly complex networks that make the brain capable of such extraordinary feats.

> In the complex system that is the nervous system, it is the excess of links within the system that allow it to fulfill its function even with far fewer links to regions that interact with the external environment.

2.4.2.3 Learning

The sheer number of interneurons in the brain means that it can be wired in a vast number of different ways. To understand how this complexity allows us to learn new skills, make sense of complex, rapidly changing situations, and come up with courses of action, we need to understand one basic principle of how the human brain learns. Consider the following:

1. A single interneuron may be connected to as many as 10,000 other neurons.
2. Imagine a child learning to read. As you've learned already, visual information is processed in the occipital lobe.
3. When a child first encounters the letters of the alphabet, the brain has no stored information that allows sense to be made of these letters. All they are to the child is a new visual stimulus.
4. When first learning how to recognize the letters of the alphabet, the child will initially be presented with an unfamiliar symbol.
5. As the visual information is processed in the occipital, the brain will identify two sloping lines that join at the top as well as a horizontal line linking them midway along their length. This, unsurprisingly, is the letter "A."
6. If you have ever seen a young child learning the letters of the alphabet, you will know that this does not tend to happen silently. It is usual for parents or teachers to sound out the letters as they are presented to the child. The associative nature of the brain means that the child quickly learns that this arrangement of slanted and horizontal lines is associated with a sound "ah."
7. Through repetition, this knowledge becomes deeply embedded in the child's memory. The symbol presented to them is no longer just an arrangement of lines but now represents the first letter of the alphabet and is associated with the "ah" sound.

8. Once the child learns to recognize such a letter, she no longer needs to analyze the arrangement of the lines every subsequent time because the neurons that are activated in the occipital lobe when the letter "A" is presented are used to being activated at the same time. This is the crucial feature that allows human beings to learn. If multiple neurons are activated simultaneously, it becomes easier for them to be activated in the future. As we said earlier, neuron activation is known as "firing." When it was discovered that it becomes increasingly easy to activate multiple neurons if they are frequently being activated simultaneously, the renowned neuropsychologist Donald Hebb summarized this concept with the phrase "cells that fire together, wire together."[2]
9. This very simple idea, when applied to different functions in the brain, explains how human learning occurs. Once the child recognizes the arrangement of lines as the letter "A," she will soon associate this arrangement of lines with a corresponding sound. Once the child has learned to recognize more letters and associate them with different sounds, she will soon associate the lines going to form "A P P L E" as corresponding to the sound of that particular word as well as to an object that can be eaten.
10. As an adult, we don't need to analyze the lines that form the letters of a particular word because we can recognize the word as a whole. Not only that, but from our experiences, if we were to see the phrase "apple pie," we don't think of two separate objects: an apple and a pie. We have learned that these words in combination refer to a specific food item, namely a pie made of apples.
11. It goes further, however. If we are used to having apple pies at home, chances are we are also used to the smell of an apple pie cooking in the oven. If we arrive home and smell that particular smell as we walk through the front door, the neurons that detect smell will recognize and remember the association between the smell and the apple pie itself. All this has been learned through association. We associate the lines with the letters, the letters with the sounds, the letters and the sounds with the word, the words with the phrase, the phrase with the object, the smell that the object, and so on. Association is key to how the brain learns and, as we shall see later, how the brain thinks. One of the best examples is when you are trying to recall a song or a poem but struggling. Once someone gives you the first line, suddenly, there is a cascade of neural network activation that summons up the next line. That line then triggers the following line, and so on.

Repeated association is the key to learning.

We will look more closely at how neurons work later in this chapter because their sheer adaptability and complexity are remarkable and will give us insights into creating and managing complex systems.

2.4.2.4 The peripheral nervous system

We need some way of getting information from the sensation detectors located within our skin to the brain, and we also need a way of getting the brain's instructions transmitted to the muscles that carry them out. The peripheral nervous system accomplishes these tasks. It is the simplest of the three parts of the nervous system and generally comprises nerves that run between the spinal cord and the body's peripheries, mainly the skin and muscles. These nerves may contain a mix of sensory and motor fibers, i.e., information is flowing in both directions through them.

2.4.3 Information processing

Let's consider how human beings interact with the world around them. At its most basic level, we sense what is happening around us (the stimuli), and then we respond to it. What happens in between sensation and response is known as information processing. To complete the picture, we tend to include sensation and response as part of information processing, although, in reality, they are the inputs and outputs of the process.

The model of human information processing that we are going to look at has five stages:

1. Sensation – absorbing sensory information
2. Attention – focusing only on that sensory information we see as currently relevant
3. Perception – forming a mental model based on the attended to sensory information
4. Decision making – deciding how we should react
5. Response – enacting our decision

To cover the mechanics of every one of these stages is beyond the scope of this book, but we will be looking more closely at the decision-making process later on in this chapter because it will be essential to our understanding of how organizations can better react to a changing environment.

> Correctly reacting to changes in the environment requires that we are able to sense the change, focus on it to form a model of what that change means, decide what to do about it, and then respond.

2.4.4 Functional scales in neurobiology

Based on what we have covered so far, it is probably no great surprise to you that the nervous system is positively fizzing with activity. With so many neurons and so many connections, something is always happening. To stratify this in some way, we need to consider the different scales at which activity is happening:

1. The nervous system as a whole
2. Within the three parts of the nervous system (CNS, ANS, and PNS)
3. Across linked networks of neurons running between these three parts
4. Across linked networks of neurons running within these individual parts
5. Within definable regions of these parts (e.g., in the thalamus of the brain)
6. In linked networks of neurons within these definable regions
7. At the synapse between individual neurons (the release, action, and breakdown of neurotransmitters)
8. At the neurotransmitter sensing or releasing membranes of an individual neuron
9. Biochemical activity within a single neuron
10. Changes to the genetic and protein activity within the nucleus of a single neuron
11. Quantum effects at the level of the atoms within a single neuron

The concept of scaling is important in how it relates to our modeling and understanding of complex systems such as the nervous system. In trying to understand some aspect of neural behavior, we must be sure that we are examining the phenomenon at the correct scale. For example, it may be enough to describe the loss of function associated with a cerebrovascular stroke by looking at the damage to definable regions of the brain. However, if we want to understand why certain psychoactive drugs work as they do, it is not enough to just look at the particular areas of the brain that they seem to target. We must ask why they are targeting that area (i.e., is there a concentration of certain subsets of neurons within that area) and then ask what are the actions of this drug on the neurons and synapses themselves. The importance of looking at the appropriate scale comes up frequently in the management of complex systems.

Understanding scales in neurobiological subsystems can also give us insight into how the structures formed that link up across different scales contribute to the observable functions of the nervous system, particularly when it comes to decision-making.

2.4.5 Decision-making

When we process information, we form a mental model of the situation and use this as the basis on which we make our decisions. This mental model is generated from sensory information and, to an extent, information from our memories that are then combined in the parietal lobe of the brain.

2.4.5.1 ACT-R

For decision-making to happen, various other capabilities are required, and there are brain regions (known as "modules" in the context of decision-making) that carry out these tasks. Their arrangement is shown in Fig. 2.4.4:

This model is known as the "Adaptive Control of Thought - Rational" (ACT-R) Cognitive Architecture and was developed primarily by John Anderson and his team at Carnegie Mellon University as a way of understanding the neural processes that underlie the phenomenon of decision-making. What follows is a description of each module and its purpose.[3]

1. Goal module

This module, located in the anterior cingulate cortex toward the center of the brain, defines and keeps track of the particular thing we are trying to achieve, e.g., replace this windscreen wiper, add 45 to 76 or ascertain what that alarm sound means. At each problem-solving stage, the goal module is referred back to ensure that the decisions being made relate to the initial goal.

2. Imaginal (mental manipulation) module

This area of the brain, located in the parietal lobe (the same area primarily responsible for building mental models), allows elements of the mental model to be manipulated to attempt to solve a particular problem. For example, it may be necessary to mentally manipulate an object to see how it fits into another object. Before learning this ability, young children can only do this by

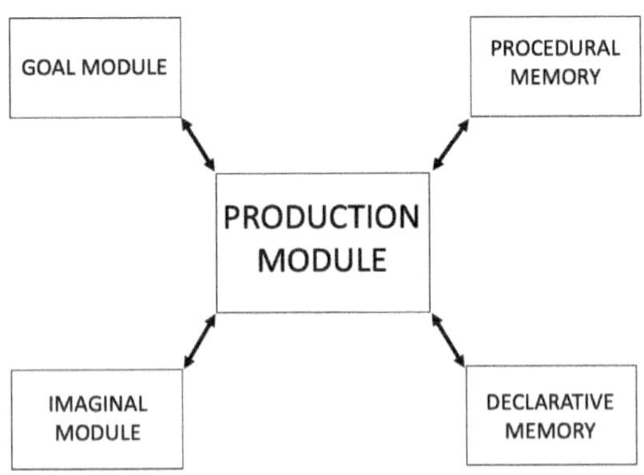

FIGURE 2.4.4 The arrangement of modules used for decision-making (part of the ACT-R cognitive architecture).

trial and error (imagine a child placing differently shaped pieces of wood into a series of differently shaped holes—there may be several failed attempts before she realizes that a square peg will not fit in a round hole. Once the imaginal module is more fully developed, the child will be able to reach this conclusion simply by manipulating the relevant pegs and holes represented in her mental model). When attempting to solve more complex problems that require several steps, the problem representation in the imaginal model is updated as each step is completed.

3. Production (pattern matching) module

This module is the main coordinator of decision-making and is thought to be localized to the basal ganglia toward the center of the brain. It interacts with the imaginal module and the goal module to define the scope of the problem. It also interrogates the memory stores to try and match the problem to past experience and so be able to select the most appropriate response. This pattern-matching process is the core of decision-making and relies on retrieving information from various memory stores to find a suitable response. The two types of memory store used are Declarative and Procedural, and the number of trials the production module is willing to go through when trying to match the best response to the presented pattern of stimuli is determined by workload.

4. Declarative Memory

Discreet bits of factual data such as the year the Second World War started, the name of your best friend at school, or the knowledge that $2 + 2 = 4$. These bits of data are called "chunks."

5. Procedural memory

This is the store of production rules used by the procedural module to solve problems. A production rule is the functional combination of neurons in a defined network that allows us to make the link between stimulus and response. They are the basis for learning and are the basic functional units used in decision-making. Neurons that have become more closely linked by repeated simultaneous firing have formed a production rule.

Production rules normally take the form IF (this problem), THEN (this solution). For example, most people will be able to solve equation $8 + A = 12$ to determine the value of A. The production rule says that IF (FIRST NUMBER PLUS A EQUALS SECOND NUMBER), THEN (A EQUALS SECOND NUMBER MINUS FIRST NUMBER). The production rule is generic, and all the brain must do is fill in the values of the first and second numbers (the "slots" in the production rule). Some problems will require the use of several production rules, and the more often they are used, the higher the chance that these multiple production rules will combine into a single, more complex production rule and so be able to be used much more quickly in the future. This is the key to learning and refers back to the material we covered earlier in the chapter regarding the importance of association in learning. Problems that could only be solved by interrogating declarative memory using simple production rules can, with practice, lead to the formation of new, comprehensive production rules. This is because, to repeat Donald Hebb's insightful words, "cells that fire together, wire together." The multiple neural firings required to activate multiple production rules lead to an association between these different networks so that the next time the stimulus is encountered, they can be quickly activated in a synchronous manner rather than individually. This ability to form new production rules allows us to develop expertise at various complex tasks.

> Production rules are how we deploy knowledge to navigate our dynamic environment. Production rules can be learned and augmented with repeated use but will eventually be forgotten or will be harder to activate if they are not activated often enough.

2.4.5.1.1 Workload

Given that neuronal communication is limited both by the speed of electrochemical transmission along the neuron itself and the chemical communication (using neurotransmitters) between neurons at the synapses, the brain has a processing speed limit that means it can only perform a certain number of tasks in a given amount of time. This limit is a lot lower than that of a computer because the transmission speeds in the brain are a lot slower than the transmission speeds in a microchip. In the same way as a computer can be made to crash or generate incorrect results in situations where the required calculations exceed the machine's computational capacity, the same thing can happen with the human brain. The brain has a certain number of tasks to perform during a fixed amount of time. It is estimated that to activate a production rule takes 50 ms, meaning about 20 production rules can be selected and tested per second. This may seem like a lot, but when you consider that humans often must perform several tasks concurrently, each needing their own production rules and that a novel task that requires pattern matching might require many of these rules to be activated and tested in the search for an adequate match, this speed limit has been described as the bottleneck of information processing.[4] Two factors can lead to this bottleneck becoming a problem, and these are the main determinants of workload. They can cause high workload when they occur separately or together:

1. High task load – multiple tasks having to be carried out within a period of time.
2. High time pressure – even a single problem, if it is complex, may generate a high workload if it must be completed within a specified time period.

Workload is a major determinant in the accuracy of decision-making.

2.4.5.1.2 Summary of the anatomy of decision-making

Given what we covered about how neurons interact with each other (i.e., neurons that are repeatedly activated at the same time form an association and so are easier to activate in the future), and given the incredible complexity and number of neuronal connections in the human brain (500 trillion), from the very basic idea of learning and practicing simple rules you can begin to see how we learn information and how our behavior, especially our amazing capacity to solve problems, is generated in the brain. Instead of one neuron being associated with another neuron, there may be a million neurons associated with a million other neurons in a highly complex way and yet still operating on the basis of that phrase that we encountered earlier—"cells that fire together, wire together." Certain subgroups of interconnecting neurons form the various modules covered in this section, and it is the interaction of all these modules that allows us to make decisions.

2.4.5.2 The two modes of human decision-making

All animals can make decisions in some form. Simpler forms of animal life can make decisions based on instinct. For example, for most organisms, there is an instinct to withdraw from painful or unpleasant stimuli. Through this simple form of associative learning, animals can learn that a certain set of environmental conditions that led to pain in a previous encounter is also likely to lead to pain if those conditions are met again. In this fashion, animals learn to avoid situations that could potentially cause harm. In the language of the ACT-R model of decision-making that we have just described, the animal is now equipped with some basic production rules that govern its behavior. Due to limitations in brain size, complexity, and available memory storage space, it is normally impossible for simpler organisms to learn more complex behaviors that require combinations of multiple production rules.

There are, however, certain production rules and other hard-wired programs that humans and some higher forms of animal life possess that allow decisions to be made rapidly under complex, uncertain, or dynamic conditions. For example, as well as being able to recognize situations that are the same as ones previously encountered, humans and higher animals can also recognize situations that are *similar* to ones previously encountered. At first glance, this may not appear to be a radical step forward in evolutionary terms, but, in reality, it is unlikely that we will encounter a situation in our environment that is absolutely identical to one which was previously encountered. The ability to judge that a situation is *similar enough* to one that has been previously experienced enables the animal to activate the same production rules as before and so execute the same successful strategy to protect itself from harm. We will return to this concept of *similarity matching* later.

Modern research in the field of decision-making has concluded that human beings possess two distinct modes that govern our decision-making.[5] In the original research, these two modes were actually referred to as systems, but as that word is already being used frequently, for simplicity, we will refer to them as modes:

1. Mode 1 – We have inherited this mode from our animal ancestors, and it allows us to make decisions in high workload situations.
2. Mode 2 – This mode has evolved more recently and allows us to make more rational decisions using all the processing power of our highly evolved cerebral cortex.

2.4.5.2.1 Mode 1 overview

Mode 1 is always active when we are conscious. It allows us to make rapid decisions when we are in complex, time-pressured, dynamic, or high workload situations. One such example would be firefighters dealing with a fire inside a building. In such a situation, there may be considerable pressure to look for survivors, concern about one's own safety, and the difficulty of dealing with a rapidly evolving situation. People who've been in situations like this often have difficulty recalling how they made the decision they did at the time. Unlike many other jobs, time pressure is a critical factor when fighting a fire, and it is rarely possible to make slow, relaxed decisions having carefully considered all the variables. And yet, decisions do get made, so there must be some wiring in our brains that allows us to do this. Of course, firefighting is not the only profession where people are faced with situations that require rapid decision-making. People working in roles within the military, healthcare, and aviation regularly encounter situations that require rapid decision-making, and one of the major research questions was how can they make decisions without conscious and deliberative assessment of the nature of the problem?

Imagine a word processor on a computer. If you pressed the letter "A" on the keyboard, you would expect it to appear on the screen. Input leads to response. What is not seen of thousands of lines of computer code that allow this to happen. For someone who doesn't know how the system is set up, trying to work out the programming that translates the push of a button into a letter appearing on a screen is almost impossible. It was the same situation for scientists trying to understand why certain conditions led to certain, apparently instinctive, and subconsciously triggered behaviors. It is only through the groundbreaking work of two scientists, Daniel Kahneman and Amos Tversky, that the unseen, subconscious programming that Mode 1 uses has been revealed.[6] The input to Mode 1 is the mental model of the situation, and the output is some sort of decision. It would seem that the brain's decision-making system is still used during this process but that it works using some pre-programmed, hard-wired production rules that we didn't have to learn but that we seem to have been born with.

The ability to make quick decisions has been passed on evolutionarily from our animal ancestors, and so Mode 1 is always operational when we are awake to allow us to react quickly to threats. When describing Mode 1, it could be called the system of instinctive decision-making. If all we had were Mode 1, we'd still be able to function as an organism, but we would behave very differently from how we behave with Mode 2 there to modify the outputs of Mode 1. If you consider the brain as a computer, the hard-wired rules that Mode 1 utilizes are like the operating system; it allows the computer to carry out some basic functions even though no more advanced software has been installed yet. The good thing about this operating system is that it seems to allow us to make lots of decisions very quickly, like the parallel processing capabilities of an advanced computer.

2.4.5.2.2 Mode 2 overview

This mode is what makes us human. If Mode 1 gives us our instincts, Mode 2 gives us our rationality. It is like our inner chess player, able to carefully assess the situation, consider the variables, calculate options and probabilities logically but also apply abstract reasoning to derive elegant solutions. The cost of this is time, effort, and energy. Trawling through memory stores, activating and testing production rules, manipulating sections of the mental model stored in the imaginal module take time. Those individual blocks of 50 ms (the time it takes to test a production rule) all start to add up. As well as that, every neuron that fires requires energy, and every impulse that is transmitted across a synapse releases chemicals that will then need to be broken down. The huge burden that Mode 2 makes on our cognitive resources can be demonstrated with a very simple experiment. When you are walking with a friend, ask her a complex, multipart question such as "start with the date you were born, add all the numbers together, square that answer and work out what 15% of that answer is." Because this task requires a lot of activity between the goal, imaginal and procedural modules, and the activation of multiple production rules, the cognitive workload is very high. What is very likely to happen is that the person will stop walking. Her brain is now so occupied with the task that it is shedding less important tasks (such as navigating the footpath).

As if the higher cognitive workload wasn't problematic enough, we may face several problems that require conscious and careful engagement, but, unlike Mode 1, we can't consciously solve multiple problems in a parallel fashion with Mode 2. Mode 2 works serially so that we can only consciously consider one problem at a time. In the mathematical example given previously, each task is carried out sequentially because the input for the next task relies on the output from the last task. If we were to be presented with a piece of paper with three unrelated mathematical problems to solve, we could only solve them one by one. These limitations of capacity and serial processing are why thinking really hard is actually very taxing, and it is why, in the words of Daniel Kahneman, "people are not accustomed to thinking hard, and are often content to trust a plausible judgment that comes to mind."[7] These "plausible judgments" are the outputs from Mode 1 but, as we'll see shortly, they can be dangerously flawed.

2.4.5.2.2.1 Summary of Mode 1 and Mode 2

Table 2.4.1 describes the different characteristics of Mode 1 and Mode 2:

2.4.5.2.2.2 How Mode 1 and Mode 2 interact and the role of workload

TABLE 2.4.1 The different characteristics of Mode 1 and Mode 2.

Mode 1	Mode 2
Subconscious decision-making	Conscious decision-making
Automatic, always operational	Controlled, needs to be activated
Low effort	High effort
Rapid	Slow
Evolutionarily old	Evolutionarily recent
Nonlogical	Logical
Parallel processing	Serial processing

Each of these modes is not used exclusively, and, in fact, they are normally both operational at the same time. When we have formed a mental model of the situation, it is Mode 1, the mode that is always operational and is automatic, that first makes a decision based on several hard-wired production rules. In high workload conditions, this decision will probably lead directly to a response, but in low workload conditions, there is an opportunity for Mode 2 to modify or override the output of Mode 1. Although Mode 1 can save us when there is no time to engage Mode 2, we tend to make much better decisions when Mode 2 has an opportunity to reconsider the problem.

Workload is the key determinant in whether Mode 2 will get a chance to modify the output of Mode 1. There is a high likelihood that the outputs from Mode 1 will lead directly to a response in any of the following three situations (or when a combination of them occurs). The first two are what determines workload:

1. High task load – multiple tasks having to be carried out within a period of time
2. High time pressure – even a single problem may be dealt with using only Mode 1 if there is significant time pressure
3. Problem underspecificity – A problem that is underspecified is one where there is not enough information to be able to make a decision based on Mode 2.

> Broadly speaking, decisions can be made using two different modes. The mode that predominates will be determined by the workload being encountered at the time.

2.4.5.3 Heuristics and biases

The hard-wired programs in our brains that allow Mode 1 to operate and so let us make quick-fire decisions are known as *heuristics*. One of the psychologists who discovered the existence of these programs, Daniel Kahneman, defines a heuristic as "a simple procedure that helps us find an adequate, but often imperfect answer, to a difficult question".[5] It is derived from the same Greek word as eureka - *heuretikos*, meaning "inventive." It can be considered a mental shortcut, i.e., a quick way of coming up with a rough-and-ready answer to a difficult question. An additional aspect of Mode 1 programming is *biases*. These biasing programs lead us to overemphasize the importance of some information, often at the cost of another piece of information. Biases can profoundly affect decision-making in a negative way.

> We have hard-wired cognitive shortcuts that allow us to make decisions quickly without using too much of our limited cognitive resources. In certain circumstances, they can be helpful, but they should be checked using normal cognitive pathways wherever possible.

Hundreds of heuristics have been identified. Many of them profoundly impact our decision-making and affect our lives in ways we don't even realize. To do justice to this fascinating area of psychological science, let's look at the evolution and classification of these heuristics.

1. Heuristics evolved because animals needed to be able to make rapid, quick-fire decisions in complex, time-pressured situations and to make

decisions in situations where all the required information isn't available, such as during social interactions.
2. These abilities needed some sort of hardwired programming that would permit rapid but approximate pattern matching of stimulus to response and for perceptual gap filling when all the required information wasn't available.
3. While these heuristics are designed to improve our performance, some of them, particularly those whose names include the word "bias," can have a negative effect on the accuracy of our decision-making because of how much our environment has changed from when we initially developed them.

Many heuristics involve replacing a complex decision with a simpler one in much the same way as we sometimes instinctively simplify problems that are presented to us so that we have to expend less effort in solving them (forced simplification). This happens when we are in situations that are too complex to understand or where the situation is changing too rapidly for the human brain to keep track of it. Remember, the human brain has evolved over millions of years. It is only in the last century that we have taken our brains, this organ that evolved to suit an organism that had a top speed of perhaps 15 miles per hour, and put them in planes and cars traveling at much higher speeds. Biologically speaking, we are still essentially prehistoric creatures. The programming in our brains is relatively unchanged from the days when our primitive ancestors walked the earth. It is no wonder that we sometimes find ourselves behind the curve. Another unfortunate side effect of the fact that the heuristic programming of Mode 1 is evolutionarily hereditary is that we can't rid ourselves of it. Many experiments have been carried out to see if people can be "debiased," but the fact is that this hard-wired programming in our brains, for better or worse, is here to stay.[8]

Aside from the specific heuristics that we will look at, there are two general patterns that some of these heuristics seem to be based on. These patterns relate to how information is stored and retrieved in the brain and how production rules are employed:

1. Similarity matching – the tendency to associate a new stimulus that is similar to one that was previously encountered, resulting in the same response. This is accounted for in the decision-making mechanisms of the brain by not requiring a perfect match between a stimulus and the appropriate response, but rather an adequate match.
2. Frequency gambling – using a strategy that frequently works in novel situations. Production rules become more easily activated the more often they are used (a process known as "raising base level activation"). In this way, a commonly used rule becomes more prominent among other similar rules and so is more likely to be activated.

There are six heuristics (some referred to as "biases") that especially have an impact on human decision-making, particularly as a cause of systematic errors that occur under specific types of uncertain or high-workload situations.

1. Confirmation bias

This is the tendency to avoid cognitive confusion by continuing to believe in a particular mental model and putting extra emphasis on environmental cues that support it while disregarding cues that would refute it. Having to abandon a mental model of a situation in light of contradictory evidence is very difficult because the brain then must build a new mental model before decisions can be made based on it.

2. Availability heuristic

This is the tendency to utilize a piece of information or a strategy that has been used recently. In a similar manner as frequency gambling, even if a production rule hasn't been used frequently, the recentness of its last usage can increase its prominence among other similar rules and so it is more likely to be activated.

3. Plan continuation bias

This evolutionarily inherited bias means that when we have settled on a plan of action, we are unlikely to change our minds, particularly when in a high workload situation. For our more primitive ancestors, when energy was scarce, after making a plan to find food, it was preferable to continue with that plan rather than switching to a different plan. Switching means that any energy expended up to that point was wasted, and so natural selection favors the tendency not to keep switching plans.

4. Representativeness heuristic

The underlying programming that leads to this heuristic is still unclear, although it does seem to be based on the idea of similarity matching. Representativeness leads to people basing their beliefs about a current situation on those from their previous experiences even though there might be salient differences.

5. Automation bias

Very simply, we tend to trust what computers tell us. To decrease our cognitive workload, we are overly willing to delegate decision-making and diagnosis to automated systems and tend to believe what those systems tell us.

6. Overconfidence heuristic

The heuristic of overconfidence needs to be clearly defined from any personality trait that might lead to overconfidence. The heuristic of overconfidence is an internal process that relates to how we assess the accuracy of our stored knowledge and our mental models. Studies have repeatedly shown that humans have more confidence in the accuracy of their knowledge and actions than is justified.[9] The ability to assess the correctness of our knowledge and actions is known as our calibration, and humans have been repeatedly shown to be overcalibrated. A normal process of feedback should lead to us refining our calibration ability, and yet this doesn't seem to be the case. As we've described already, this heuristic of overconfidence biases our own judgment, and heuristic based biases cannot be overridden through training. The evolutionary basis for this is intriguing. For more advanced organisms, an important social ability is being able to detect deception in others. For a social organism that is intent on convincing other members of its species that it is more dominant, strong, or intelligent than it really is, the most reliable way of doing this without being detected is to deceive itself as well, i.e., to genuinely believe this to be the truth. This is the reason that we are more confident in our abilities and knowledge than is justified.

> Our hard-wired heuristics and biases can affect the accuracy of our decision-making in a consistent way. It is only by actively engaging Mode 2 whenever possible that we can detect these potentially inaccurate responses before implementing them.

2.4.5.4 Algorithmic decision-making

I want to introduce this idea here, even though it will become a lot more relevant in Part 3 of this book. In 1954, a psychologist at the University of Minnesota, Paul Meehl, wrote a book entitled "Clinical versus statistical prediction: a theoretical analysis and a review of the evidence."[10] This book was remarkable because, for the first time, it suggested that psychiatric diagnoses that were arrived at in an algorithmic (i.e., statistical) manner were generally more accurate than clinical diagnoses. In practice, the algorithmic approach involved comparing a patient's symptoms against a list of potential diagnoses rather than trying to arrive at a diagnosis through a clinical interview. In the first scenario, a psychiatrist need not even be involved in the process, and it would be possible for a layperson to use the algorithm to arrive at an answer. When the patients were followed up, it was found that the diagnoses were more likely to be correct for those who were diagnosed via the algorithmic system. The reason it is worth considering this phenomenon here is because algorithmic decision-making can make up for some of the shortfalls which are inbuilt into the human decision-making mechanism.

Paul Meehl's insight has proven true across a wide number of fields, from wine growing to sports teams, from stock picking to law enforcement.[11] An area that is particularly interesting but still potentially controversial links back to what first impelled Paul Meehl to write his book: diagnostic medicine. If you were to put a patient with a certain set of symptoms in front of ten different doctors, you would potentially get a range of different possible diagnoses. While it is possible that all these diagnoses may be wrong, it is not possible for all these separate diagnoses to be right. Doctor A may have just finished a course on acute intermittent porphyria, and because the production rules associated with a diagnosis of acute intermittent porphyria are higher up in her neural network, she begins to see acute intermittent porphyria everywhere.

Doctor B may have just watched an episode of a medical drama on television and, subconsciously wanting to make a similarly amazing diagnosis as the on-screen doctor did, comes up with a diagnosis of abdominal epilepsy even though it is incredibly rare.

Doctor C may assess the patient and determine that the patient's pain is caused by trapped intestinal gas.

Although this is a contrived example, it is true to say that doctors reach their diagnoses in different ways. The benefit of an algorithmic approach is that for different patients who are showing a particular set of symptoms, the algorithm is going to give suggestions that are unbiased by recent experience or knowledge level. An algorithm can account for differences in biological sex, age, gender, and ethnicity in its calculations in a reliable and predictable way. It may not be able to pin down the diagnosis in the first instance, but it could certainly suggest the next steps for further investigation to narrow down the list of potential options.

The reason algorithmic decision-making is very controversial is that doctors are understandably quite protective of their right and ability to diagnose patients. The idea that an algorithm could do it more quickly and reliably is not popular. However, when reviewing the evidence for algorithmic versus clinical diagnosis, algorithmic decision-making comes out ahead.[12] When it comes to managing complex systems, we need to factor in the reliability of algorithmic decision-making into our management strategy.

> Algorithmic decision-making has repeatedly been shown to give more reliable and accurate responses than subjective decision-making. This is most likely due to the impact of heuristics and biases on subjective decision-making.

2.4.6 Levels of performance

In a broader sense, there is more to human behavior than just decision-making. Another way of looking at how humans interact with the world is to consider how deeply they are engaged in the process that leads to a particular behavior. In 1983, Jens Rasmussen proposed that humans perform at three different levels.[13] The term "performance" refers to some sort of interaction with an external element and, importantly, is often associated with attempting to solve some sort of problem. Performance levels are relevant because humans don't always operate flawlessly, and failure can occur at any level of performance. There are distinct characteristics in the types of errors that occur, the rates at which they occur, and, more importantly, the rates at which they are detected and adequately corrected. By understanding how humans do and don't perform successfully when operating at these different performance levels, we can begin to understand how humans may become a point source for failure within a system. We will look at the specifics of these potential failure modes shortly. The three performance levels in humans are as follows:

- skill-based (SB) – Performance is preprogrammed and automatic so that the individual does not have to engage consciously with the problem to react to it correctly. For example, when driving a car, if it diverges from the center of the lane, it is instinctive for the driver to correct this without having to think about it consciously.
- Rule-based (RB) – Performance at this level is semiautomatic and does require more conscious engagement than the SB level. To perform at the rule-based level, patterns are identified, and these are associated with production rules stored in the form of "IF (this pattern), THEN (this response)." For example, once a driver has been trained, she will know that certain street signs mean she should respond in a particular way. She will be consciously aware of the sign, and, because of her training, she will have a stored production rule that says, "IF this sign is seen, THEN respond in this way." She can then activate this rule and respond quickly without having to consider the meaning of each sign. It is worth noting that human performance at the SB level is actually based on rules as well, but these rules have become so well-practiced and have been used so many times that they now become skills as they don't require conscious activation or monitoring. If you consider the example of learning to drive, it is initially difficult as rules regarding gear sequences and clutch use need to be consciously activated. Eventually, these rules become so well-practiced that they can be activated without conscious engagement, and we can now perform many driving tasks at the SB level.
- Knowledge-based (KB) – This level of performance occurs when the individual finds no suitable rule for determining what should be done in a particular situation. The only option is to examine the individual's store of knowledge in an attempt to understand what is going on. This is a conscious, high workload process and requires selective attention on the problem. Suppose the problem is adequately solved using this knowledge-based process. In that case, it is highly likely that the original pattern of stimulation will be matched to the final response, and the person will create a new production rule that will enable them to solve the same problem using rule-based problem-solving should they encounter it again in the

future. This, in principle, is learning; through our experience, we store more and more pattern-response production rules that allow us to solve problems more rapidly and accurately as our experience grows. In fact, the psychologist James Reason defines "expertise" as never having to rely on knowledge-based problem solving, i.e., as the person has more experiences and encounters different scenarios, many of which will require KB problem solving, this KB problem solving will result in the production and storage of many more production rules.[14] This means that the highly experienced person will be able to call on these production rules in the future and will now be able to handle the vast majority of situations at the RB level and the need to resort to KB processing is increasingly unlikely unless a truly novel situation is encountered.

> The ease with which we generate responses to stimuli depend on which level of processing we engage during specific tasks.

2.4.6.1 Human failure modes

Before we get to the part of the book where we try and apply the lessons and concepts from this part of the book to the task of designing better systems and managing them more effectively, we are going to have a look at some situations where complex systems were exposed to some sort of disturbance. This could have originated from within the system or been generated externally. In some cases, the system manages to deal with these disturbances without a loss of function. In other cases, the system collapses completely.

Despite most of this book being about how things go wrong on the systemic scale, we need to consider also how things go wrong at the level of components. We've already said that humans are a component of socio-technical systems. We know that human performance varies, and this variability can be the source of disturbances generated within the system. Therefore, it would be sensible to consider how and why individual humans make errors or make poor decisions. Often, there will be systemic contributory factors that lead back into reexamining the system as a whole, but, for completeness, we briefly need to focus on the actual moment when the human component of a system makes some sort of mistake that potentially sets a disturbance in motion.

Before we discuss how errors can occur at each performance level, we need to get into the details of what we mean by "error." Let's look at a commonly used definition: error will be taken as a generic term to encompass all those occasions in which a planned sequence of mental or physical activities fails to achieve its intended outcome, and when these failures cannot be attributed to the intervention of some chance agency."[14] To simplify this, we can say that error is defined as "the failure of planned actions to achieve their desired goal."

Having discussed the three levels of human performance, we can start to divide this definition further. If an error is the failure of planned actions to achieve their desired goal, there are two ways that this failure can happen; either the plan is wrong, in which case our actions are doomed to failure from the start, or our plan is correct, but our execution of it fails for some reason. With this in mind, let's look at the types of error that occur at the skill-based (SB), rule-based (RB), and knowledge-based (KB) performance levels as laid out by James Reason in his book, "Human Error"[14]:

1. Errors at the SB level – slips and lapses—Our plan of action, although it has been selected automatically without conscious effort, is correct and suitable. The error occurs during the execution of the plan. Inattention is a classic reason for this, e.g., intending to drive to a location near your workplace and finding that you have driven to work instead. Another example would be forgetting to do something due to interruptions or distractions.
2. Errors at the RB level – The production rule we have selected to deal with the current problem is not appropriate, i.e., the plan is wrong. As you will see in the next paragraph, if we consciously recognize that we can't solve a problem at the RB level, we should move up to the KB level and try and analyze the problem further and search our knowledge stores to be able to start solving it. If this is what should happen, why then might we not move to the KB level and, instead, stay at the RB level and try to use a rule that turns out to be unsuitable? The answer can be found in our earlier descriptions of RB and KB performance. KB performance takes a lot of conscious effort. It requires selective attention to be focused on the problem and an effortful reevaluation of the problem parameters followed by a slow trawl through our knowledge stores to try and develop some sort of plan to respond to this novel situation.

The reality of many of these novel situations is that we don't have the luxury of spending time and cognitive resources coming up with the best possible answer. There may be hundreds of other situational factors calling out for our attention. We can't always take each problem and devote the time and resources to it in the hope of getting the best possible answer. High workload situations make it much more likely that we will select a production rule to use for a particular problem that we think *might* be suitable or that we think will give an adequate solution rather than the perfect solution. In selecting these "rough-and-ready" but quickly accessible production rules, our brain is programmed to take several heuristics into account, the most powerful of which have been considered in the context of Mode 1 decision making but are worth reviewing here:

   a. Similarity matching – while the pattern we are dealing with is novel and there is no production rule that we can use to come up with a response, is the pattern *similar enough* to another pattern that we have encountered where we know that we have a production rule that we can use? The reality is that we may overemphasize the similarity between two situations and so miss crucial differences that would mean that the production rule we're planning to use isn't going to work.
   b. Frequency gambling – A frequently used rule is more likely to be used than an infrequently used rule. Let us say that we use similarity matching to decide that a novel situation is similar to two situational patterns that we have encountered previously. We are then likely to use frequency gambling to decide which of the two situational pattern's corresponding production rules we use, i.e., the rule we've used most often is the one we are likely to pick. The reality is that we have engaged in a statistical act of risk taking. We are gambling that a rule that is used frequently will have a higher chance of addressing the problem. This is not always the case.
   c. Overconfidence – Studies have repeatedly shown that we have more confidence in the accuracy of our knowledge and abilities than is justified.[8] Human beings are generally more confident in their judgment of the correctness and suitability of their actions than is justified. Overconfidence reinforces the similarity matching and frequency gambling programming that we have and increases the likelihood of making an RB error.

3. Errors at the KB level – provided the workload is low, we may consciously recognize that a novel problem can't be solved at the RB level. Error can occur at the KB level when attempting to derive a solution by analyzing the problem more deeply, followed by conscious interrogation of knowledge stores to understand the situation and come up with a solution. The four main causes of KB errors are:
   a. Missing knowledge – a student taking a French translation exam may try to understand a phrase that she hasn't encountered before by looking at the context or trying to derive its meaning from similar phrases. However, if the required knowledge isn't in her memory, there is a high risk of making a KB error. She simply doesn't know what it means.
   b. Processing limitations – it may be that the situation is evolving so quickly that the individual can't maintain an accurate mental model on which to base their decisions. This isn't uncommon when trying to solve problems occurring in highly automated, highly interconnected sociotechnical systems such as financial institutions or nuclear power stations.
   c. Oversimplification – a strategy that we sometimes employ in the face of highly dynamic situations is that we may try and replace a complex

problem that is seemingly too difficult with a simpler one. The risk is that in trying to eliminate complexity, we ignore some vital elements of the problem and so come up with an unsuitable plan.

d. Overconfidence – in the same way as with RB errors, we generally have more confidence in the accuracy of our skills and knowledge than is justified, and this will reinforce any KB solutions we come up with even though they may be wrong.

Errors do occur in complex systems. Humans generally try and do their best at work, but that doesn't always work out as expected. In some of the case studies you will see later, an error may have triggered the event. Given the complexity and interconnectedness of the systems within which we work, we know that these errors don't just occur spontaneously. However, by figuring out what type of act it was and what performance level the person was using when they performed this act, we might get clues about weaknesses in the system that allowed this act to occur.

> Human components in the system can perform acts that cause wider disturbances, and understanding the different causal basis of a particular act can assist us in reducing the chances of it happening again in the future.

2.4.7 Neural dynamics and connectomics

If you wanted to simulate the function of a human brain in a computer, how much detail would you need? Would something based on the ACT-R model be enough? Evidence shows that it does a good job of simulating brain function for relatively simple tasks.[15] The question is, what size of computer would you need to replicate everything, i.e., "sensory" information of any description can be fed into the program, and the response would be that of a human. More specifically, it would be that of a particular human, i.e., the human whose brain you had used as a model for the simulation. If that human liked pistachio ice cream, how much detail would be needed to make the program want that ice cream when it "saw" it, and how would the memories of the childhood holidays during which the human whose brain was being simulated first had pistachio ice cream be evoked?

If the brain were planet Earth, then a single neuron would be a person. We can say that the brain itself is a complex system because the individual components, the neurons, demonstrate such high levels of adaptability. Because neurons can have so many connections with other neurons and because they can change the way they express receptors to a wide multiplicity of different neurotransmitters, we can say that neurons in themselves are highly adaptable. We can also ascribe a quality to them which isn't really relevant for other cell types in the body. Because of their unique internal dynamics and their unique connections, neurons have an *identity*, i.e., we cannot simply swap out one neuron in a network for another. It would be like replacing a highly capable, highly experienced team member with a brand-new recruit and expecting the team to perform exactly the same way.

We can also say that neurons, certainly at the local level, are highly coupled (to use one of the descriptions of complexity defined by Charles Perrow). Imagine a human being who had the capability of affecting 10,000 other human beings in a very real and immediate way. The 500 trillion highly variable connections (the synapses) give the brain its capability. To use Ross Ashby's term, they give the brain its requisite variety.

In an earlier part of the book, we looked at the need for complex systems and specifically Ross Ashby's Law of Requisite Variety:

> The more we expect our systems to do, and the more resilient we expect them to be in the face of disturbances exerted by the dynamic environment in which they function, the more complex the system itself must be.

The human nervous system is a perfect paradigm for this law. When we consider the 86 billion neurons in the human brain, in combination with approximately 500 trillion potential synapses at each one of which any or several of up to 40 known neurotransmitters act, we start to get an idea of the sheer number of potential states that the human nervous system can be in. Even if we focused purely on the interconnections between neurons, referred to as the human "connectome", the number of potential combinations is staggering.[16] When we compare this number to the number of potential environments that the human being might find themselves in, we can begin to see how the complexity of the system designed to help us operate within these environments more than matches the complexity of the environment itself.

Some fascinating work looking at how difficult it would be to truly simulate the functions of a human brain suggests that even if we were just to try and replicate the human connectome, i.e., the complete map of the connections between different neurons throughout the brain, we do not possess the technology.[17] To calculate the number of potential states that the simulated connectome could be in would require computing power that is only forecast to be developed in 50 years, based on the current predicted rate of development of computer technology. Bear in mind, modeling the human connectome is the most basic meaningful way of simulating the human brain. It is likely that this simulation would be relatively crude, and to more accurately represent the functions and dynamics of the human brain, computing power would be required to map and process the activity of individual proteins within neurons and at synaptic connections. The computer technology required to do this may not be available for 100 years, if ever.

It is fair to say that the human nervous system is the most complex entity in the known universe. If we currently don't have the capability to be able to simulate the human brain, perhaps some of the recent insights from neurobiological research might give us a way of accessing deeper knowledge about how it performs its myriad functions.

An individual neuron itself is highly adaptive. It stands to reason then that the networks formed when multiple neurons, perhaps in the region of hundreds or thousands, come together to form a neural network make the resulting system (network) highly adaptive.

In the same way as we discussed entropy when it comes to the compartmentalization of information, neurons that form these connections are essentially compartmentalized from those neurons that are not wired into that same network. To maintain this compartmentalization against the effects of entropy, energy, in the way of periodic reactivation, is required to reinforce these connections. Evidence suggests that as a first step, new neural networks and connections formed during the day are reinforced as we sleep.[18] The changes in brainwaves observed when people sleep suggest that the networks formed during the day are being subtly reactivated to reinforce their connections. However, anyone who has had experience learning a language at school but then doesn't practice it frequently throughout their lives will find that those connections that were formed at school have diminished over time. It may be possible to reactivate these connections when trying to string a sentence together in a different language, but we may find that the connections required to retrieve a particular word in that language from our memory stores are no longer there. There may be a remnant of this connection, which means that when we come to try and learn the same language for a second time, we find it easier than the first time, but this again goes to demonstrate how constant reinforcement of neural networks is required to maintain knowledge. This is true for the central nervous system but less so for the autonomic nervous system.

We have certain capabilities that are hard-wired into the structure of the autonomic nervous system which means that we do not have to learn or remember how to make our hearts beat or how to breathe. There are rare neurological cases where damage to a specific brain area can knock out one of these hard-wired capabilities. For example, damage to one particular region means that the person then must concentrate consciously on breathing in and out. These autonomic functions are inherited from our mammalian and reptilian forebears in our evolutionary line and represent the functions required to sustain life.

In summary, we can say that neurons are highly adaptive but that the capabilities of the human nervous system lie in large part with the nature of the connections between neurons (the connectome). We also know that we have some hard-wired neural networks in addition to the more pliable and transient connections that we form ourselves through learning. As we have seen, these two types of neural networks come into play in decision-making, particularly when we are under pressure.

> It is the flexible, adaptable network of connections between neurons that go to make up the human connectome that endows us with the ability to anticipate, monitor, respond and learn from our interactions with our environment. In addition, we have certain hardwired capabilities that mean that the essential processes of our body are regulated without conscious cognitive effort.

2.4.8 Brain plasticity

Human connectomes are so complex as to be unique compared to other connectomes; no brain is alike. It is also fair to say that even within a single human connectome, such is the degree of possible variation in terms of synaptic connections and metabolic activity, that every neuron is unique. If a neuron is unique, then every neural network is unique, and every networked functional brain area is unique. How can the human survive when something happens that damages one of these unique structures?

We are fortunate that not all brain regions are of equal importance. Most sensory and motor functions on the left side of the body are processed in the right-hand hemisphere of the brain and vice versa. However, how does that logic work for language?

In most people, the brain's language centers are in the left hemisphere. Damage to a particular part of the right hemisphere may have very little effect, but if that damage was to the corresponding area in the left hemisphere, the person could lose all their language abilities, both speaking and understanding.

As anyone who has had a friend or relative who has experienced a cerebrovascular stroke, the initial deficits that such a stroke can cause can appear to be devastating. However, with proper rehabilitation, the brain can do a certain amount of functional redistribution and shift certain functions to other brain regions.

We tend to lose this ability the older we get. Children have a much better prospect of being able to rewire their brains if parts become damaged or must be removed because of a particular illness. In some cases, children have had one entire half of their brain removed and, after some rehabilitation, have no obvious functional deficits, and grow up without any neurological problems.

> In the event of damage, it may be possible to redistribute functions to other brain areas.

2.4.9 The autonomic nervous system

Keeping the human body alive in a dynamically changing world is a delicate, complex balancing act. Not only do we have to adapt to our environments, but we also must adapt to the demands that we are placing on our bodies at any one time. Fortunately, our autonomic nervous system does an excellent job of keeping us going.

2.4.9.1 Homeostasis

For a wide range of biological parameters, the body can only function if these are kept within a specific range. These parameters include body temperature, blood pH, oxygen levels, heart rate, breathing rate, and blood sugar levels, to name but a few. The ability to keep these parameters within acceptable levels is known as homeostasis. For example, in the case of body temperature, we know that the human body can only tolerate a relatively narrow range. If our body temperature strays outside of this narrow range, the normal functions of our body become impaired until we reach the point where our biochemical processes are no longer able to continue.

The autonomic nervous system can employ various measures to keep our bodies within an acceptable temperature range. For example, if it detects that body temperature is too low, it causes the muscles to start shivering to generate heat. If it detects that body temperature is too high, we begin to sweat so that the evaporation of the sweat will lead to a decrease in temperature. Similar processes keep all the other essential biological parameters within their optimum range. All of these homeostatic mechanisms are based on the principle of negative feedback: adjusting the physiology of the body to minimize the gap between where a particular parameter is and where it should be.

The principle of homeostasis is about keeping certain biological parameters stable. Sometimes, however, homeostasis is not adequate to allow the body to continue to function adequately. It could be that our environment is exerting such continuous pressure on us that simply trying to maintain the status quo is going to be a futile process. Instead, the body must adapt over time to these new outside conditions. This process is known as allostasis.

2.4.9.2 Allostasis

Allostasis is the ability to maintain stability through change.[19] Allostatic models of illness consider the inputs (stressors), processing (subjective psychological assessment of stressors), and outputs (stress response processes). It can be differentiated from homeostasis by the timescales involved. Humans have homeostatic mechanisms that keep blood glucose levels, blood pH, and a myriad of other things controlled within a narrow range by making short-term alterations to the body that are easily and completely reversible. An allostatic mechanism changes the body's biology to achieve stability (or near-stability) when a stressor is present for a prolonged period. These allostatic mechanisms may involve the production of stress hormones, alteration of brain function or gene expression, or other physical changes in response to the stressor.

Under normal circumstances, once the stressor is no longer present or the individual has adapted sufficiently to cope with it, these allostatic responses decrease. For example, allostasis comes into play when someone begins to suffer from chronic insomnia. To achieve some sort of stability in the face of chronic, performance-degrading cumulative sleep deprivation, the allostatic response is to activate the hypothalamo-pituitary axis (the brain's stress response system) to start producing stress hormones to give the individual enough chemical stimulation to get through the day. The problem is, unlike homeostasis, which is short-term and reversible, allostasis comes at a price, especially if allostatic mechanisms are repeatedly called upon. For the chronic insomniac, repeated activation of the hypothalamo-pituitary axis leads to a build-up of chemical stimulants which persist over time and make it even more difficult to get back into a proper sleeping pattern. This is an example of a pathological positive feedback loop (see next section).

Allostatic load is the "wear-and-tear" that the body experiences because of chronic stress. Allostatic mechanisms are designed to allow the person to maintain stability through change as a response to a short-term stressor. A significant life event may cause an increase in stress hormones and an alteration of brain function, possibly even short-term depression, but over time, these stress responses will reduce until the individual is back to normal. However, there are four allostatic states that can lead to excessive allostatic load and a high risk of illness[20]:

1. Repeated hits – repeated experiences of stressors, often different ones that don't allow for easy adaptation, e.g., you experience several different stressful life events one after the other.
2. Lack of adaptation – this is like "repeated hits" except it is more likely that the stressor is the same. Unlike the example given above, if an individual experiences the same stressful life event repeatedly, e.g., losing one job after another, there is an opportunity to develop a coping strategy based on this repeated experience. In essence, there should be some sort of adaptation to chronic, repeated stressors, and if coping mechanisms aren't used, there will be the same allostatic response to a stressor rather than a slowly decreasing response that would be associated with learning how to manage this stress in a less damaging way.
3. Prolonged response – the response to the stress persists long after the stress has passed, e.g., posttraumatic stress disorder.
4. Inadequate response – the response to the perceived stressor is inadequate, meaning that allostasis isn't actually happening, and so there is no

adaptation. The effects of the stressor persist and are met with a low-level response that persists.

All these states can lead to excessive allostatic load because the stress response persists. The allostatic load can be measured as a function of various biomarkers, including stress hormones, metabolic markers (e.g., cholesterol and lipoproteins), immune system markers, and cardiovascular measures (e.g., blood pressure). Once these markers are combined, the results are closely correlated with impaired cognitive and physical performance, decreasing performance, and a significant risk of disease which will affect the functioning of the biological system in question.

Allostasis will be an important feature of the framework that we will build in Part 3 as it represents an important capability of a system—stability through change. How systems deploy allostasis successfully, or unsuccessfully, can determine their viability. As we can see in biological organisms, the relationship between a stressor (in the language of systems, a disturbance) and the biological reaction to it will determine if the organism can adapt successfully or not.

Homeostasis and allostasis keep biological parameters stable and, when necessary, change those parameters' set-point to handle persistent external disturbances.

2.4.10 Feedback

In Part 1, we saw an example of feedback in Norbert Wiener's design for a gun that could track an aircraft and progressively decrease the error between the shell's destination and the aircraft's position. This design was based on the principle of negative feedback. Negative feedback is also essential for many biological processes, especially homeostasis.

In a system, negative feedback works by activating some sort of process when the value of a parameter falls outside a specified range, e.g., the thermostat in a house switches on the heating when it detects that the temperature has dropped outside the set constraints and switches it off again once the temperature is restored to the set level. In a biological system, chemical receptors detect the level of sugar in our blood. If the level falls too much, the body starts to convert glycogen stored in the muscles and liver into glucose. If blood sugar is too high, it converts it into glycogen to be stored and used later. These processes are based on negative feedback because they switch off when blood sugar levels return to within normal limits.

Positive feedback is the opposite. When a positive feedback process is underway, it amplifies the process in question until some sort of functional limit is reached. A microphone held close to a speaker will pick up noise coming from the speaker, feed it back into the speaker, which then amplifies it to be picked up again by the microphone, and again, and again in a cycle until the volume exceeds the speaker's ability to produce it. At this point, the process stops. In the human body, blood clotting after an injury works by positive feedback. Through a complicated biochemical process, the early steps in the clotting process make it more likely that further clotting will occur, and then these biochemical steps also promote the clotting process until a structurally complete clot is formed.

It is easy to imagine that most processes in the uninjured human body are controlled through negative feedback. Viability and stability tend to go together. Positive feedback can occur as the result of injury in an advantageous way, but it can also occur in disease states in a way that can potentially be harmful. In the case of an allergic reaction that is so severe that it becomes anaphylactic shock, dysregulation of the immune system means that every mounting step of the immune response amplifies the subsequent step. When this happens, the body's reaction after exposure to a relatively harmless allergen spirals into a life-threatening overreaction of the immune system mediated by a positive feedback loop. It may only be the patient's death that brings the reaction to an end.

> Feedback loops underpin some of the essential dynamics within the brain and body. Negative feedback loops are prevalent in normal homeostatic control processes. Positive feedback loops are more likely to be pathological and potentially damaging.

2.4.11 The reticular activating system

The reticular activating system (RAS) is a small but important part of the brain. It is a cluster of neurons, approximately the width of a pencil and measuring a few centimeters in length within the brainstem. While most brain regions have a clear function, the RAS is a little bit more general in what it does.

The RAS is essential for managing our level of consciousness and regulating the transition between sleep and wakefulness. It has some crossover functions with the autonomic nervous system, specifically with regard to breathing and heart rate, and is also highly active when we dream. One of its most interesting functions is how it impacts our attention level, also known as our arousal level. We are regularly bombarded with more sensory information than our brain would be able to process. The RAS acts as a gatekeeper between the flow of sensory information rushing upward through our spinal cord and the higher cortical processing capacity that we would need to deal with it. It does this by filtering out the unnecessary sensory information, which it thinks we do not need to pay attention to. To do this successfully, it needs to know what we *do* want to pay attention to. To set this "filter," the RAS and the rest of the higher cortex need to coordinate to a certain extent to ensure that important sensory information is passed upward into the brain for processing while less important information is filtered out.

Under normal circumstances, the brain detects external stimuli as information is collected through the sensory organs in the skin and is then transmitted through the peripheral nervous system into the spinal cord and upward into the brain. Rather than giving us specific information about sensory stimuli, the RAS gives us more of a general sense of what is going on around us, i.e., it tells us something is happening but not necessarily what that something is. This function is active whether we are asleep or whether we are awake. For example, when we are sleeping, the activity in our cerebral cortex is much reduced. However, it is still possible for us to be woken by an external stimulus even though we might not necessarily remember what that stimulus was once we have awoken. It could be a loud noise, or it could be excessive heat or cold, but it is the reticular activating system rather than our normal sensory pathways which have detected that something is going on in our immediate environment and has gradually increased our level of consciousness in case it is something that we need to deal with.

> Even when we are unconscious, we continue to monitor our environment.

2.4.12 Reflexes

Although the central nervous system can perform a startling array of functions, there are many bodily functions and responses to stimuli that can happen without the involvement of the CNS. These are known as reflexes.

Some readers may have had occasion to have their patella reflex checked. This normally is part of a medical assessment of the nervous system and involves sitting with the lower leg hanging loosely while someone taps the tendon that runs from underneath the kneecap. People have differing levels of response to this, but in most cases, the tapping of this tendon causes a reflex in the quadriceps muscle of the thigh, and the lower leg kicks out slightly. Stimulus has led to a response, and, in normal circumstances, the normal process of quadriceps activation following tendon stretch is an important part of maintaining our balance without having to think about it constantly.

Interestingly, this process doesn't follow the normal pathway that we described earlier in the chapter (sensation-attention-perception-decision-response). Although we can feel what is going on, we do not consciously initiate the response. Instead, the stimulus is directly linked to a programmed response within the spinal cord.

The withdrawal reflex is another reflex over which we don't exert conscious control. A sudden, unexpected painful stimulus, such as touching a hot object, results in us instinctively withdrawing our hand. Again, this protective reflex is mediated by the spinal cord rather than our higher cortical functions.

In some cases, we can overcome certain reflexes if we are ready for the stimulus, but these spinal reflexes exist so that we can offload a lot of the unnecessary processing that would otherwise have to be carried out by the brain. It doesn't take a genius to know that you should take your hand off the hot surface!

Reflexes are important to consider when we look at the overall nervous system because it can be easy to assume that our higher brain functions are involved in every aspect of our behavior. The truth is a little bit more complicated, and when it comes to reflexes, levels of arousal, and autonomic control, we can see how control is, to an extent, decentralized throughout the body.

> Reflexes are a way that essential stimulus/response functions are decentralized so that they can occur quickly and without conscious effort.

Chapter 2.5

Sociotechnical systems

Chapter outline

Any customer can have a car painted any color that he wants so long as it is black.

Henry Ford, maker of the first production automobile

2.5.1 Introduction

A sociotechnical system is a class of systems that best represents human endeavor and is based on how social and technical systems intersect. They were first formally described in the early part of the 20th century when researchers in England were analyzing the organization of coal mines. Realizing that the human operator and the tools and machines that helped them achieve their objectives were tightly connected, they concluded that the interrelatedness of the workers, the machines, and the rules governing operations combined to form this particular type of system. The term "sociotechnical" remains widely used in industrial engineering.[1]

During the 1970s, this idea of a sociotechnical system led to the development of the SHELL model. The SHELL model, SHELL being the acronym of its four components, considers all the components of a sociotechnical system and is shown[2] (Fig. 2.5.1):

The "environment" component is self-explanatory; the liveware refers to the human components, hardware refers to the machine components, and software refers to any procedures, software, or laws (i.e., the informational

Designing and Managing Complex Systems. https://doi.org/10.1016/B978-0-323-91609-7.00006-8

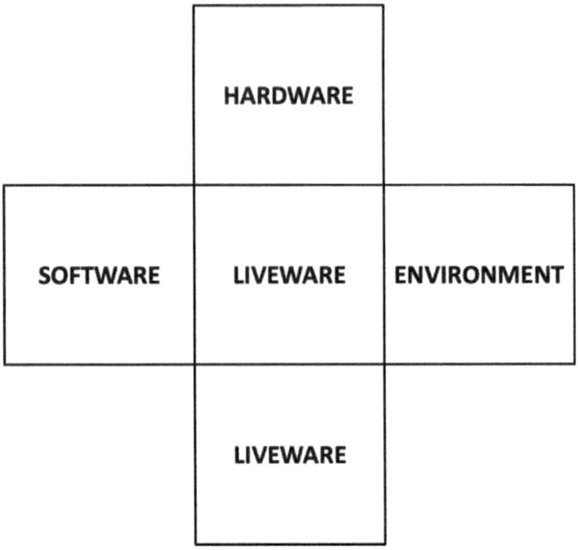

FIGURE 2.5.1 The SHELL model.

system) that impact the operation. As you can see from the diagram, there is liveware at the center of the model. This liveware is the human operator in the system from whose perspective we are viewing the rest of the system. That human operator will need to interact with machines (hardware), with procedures (software), and with other people (liveware). All of this is done in some sort of environment. Aside from the individual components, the real insight that the original designers of this model had was to focus on the interfaces between elements, i.e., the points where the squares touch. In trying to improve a sociotechnical system like this, as well as wanting to optimize the components themselves, we also need to consider how we can optimize the *interfaces* between components. For example, the interface between liveware and liveware relates to how successfully two (or more) human operators in a system work together, and optimization of this interface will be about helping these human operators to more accurately share mental models, work together when needed, and show reciprocity when circumstances require it.

What also distinguishes sociotechnical systems from other classes of systems is that they are normally created for some specific purpose. As we have seen, sociotechnical systems are common in areas of human endeavor, and this is particularly true of commercial organizations. We can say that the function of a system is its purpose.

Going back to the example that we used previously in the book, a commercial airline has the following function: the safe, efficient, and profitable transport of people and cargo between airports. This is a clearly defined function, although, as we will see later, trade-offs may need to be made between safety, efficiency, and profitability. However, once we have struck this balance, the job of the manager is to try and optimize the components within the system (and, crucially, the interfaces between the components) to achieve this function. Part 3 of this book will be dedicated to how we can best arrange components and ensure that the interfaces between them are as efficient as we can make them.

> As well as optimizing components, we need to optimize the interfaces between components.

2.5.2 Scales in sociotechnical systems

Although we have talked about scales in the context of systems in general, from here on in, you are going to start seeing another word coming up more and more frequently. That word is recursion.

If we take an organization as a whole as our starting point, a media company, for example, we can ascribe a certain pattern of organization to the whole enterprise. Most organizations have a corporate head, internal channels for communication and audit, and, ultimately, "front-line" operational teams that interact with the environment outside of the organization to fulfill the function of the media company. The whole media company is a system.

Let us now look at one part of that system, i.e., one of its subsystems, for example, the television news division. Although this is a subsystem, it will likely have a similar organizational structure to the organization as a whole; there will be a few people in charge, there will be channels for communication, audit, and distribution of resources, and then there will be the front-line operational departments who get the job done, i.e., the various teams of journalists working on specific areas of the news (sports, economy, international affairs, etc.), the researchers, the support teams, etc.

We can go one step further down and look at the department that specializes in arranging the logistics for the international reporting team. As you can guess, the general organizational structure is repeated until we get down to the final operational unit, a human being arranging the transportation of equipment to an international destination.

The final step, and the one that forms the basis for Part 3 of this book, is the human. As we have introduced in the chapter on neurobiology and as we will see in a lot more detail later, the general arrangement of a human, with specific regards to their nervous system, again follows this organizational pattern.

The take-home message is this: although we can zoom in and out to different scales of an organization, the general organizational arrangement will remain the same, i.e., the arrangement recurs within subsystems, within their subsystems, within their subsystems, and so on down to the level of the human ... where it recurs again! To complicate matters further, we must also remember the effect of scaling when we consider the Cynefin domain that systems are operating in. Just because the news division seems to be operating in the Complicated domain, it doesn't mean that some of its subsystems aren't operating in the Complex domain (consider a news unit embedded in a war zone).

Overall, recursion is a useful idea to embrace because it means that by defining some principles about how to best manage a sociotechnical system as a whole, we can apply those same principles right down to the level of the human working on the front line. Recursion is there at every level, and it makes life easier for us.

> Successful management of a sociotechnical systems requires us to view it at the correct scale.

> Recursion at different scales across sociotechnical systems can be useful because it suggests that management techniques that work on one scale might work just as well in an embedded recursive subsystem.

2.5.3 Taylorism, Fordism, and requisite metasystems

Frederick Winslow Taylor could be described as the first management consultant in history. Working during the latter part of the 19th century, he introduced the idea of "scientific" management. By analyzing the workflows within various organizations, usually factory-based ones, he developed techniques that would improve efficiency across a range of sociotechnical systems. It was Taylor who was credited with founding the field of industrial engineering.[3]

For those of us who work in modern sociotechnical systems, where, for the most part, there is a general understanding of organizational efficiency, it is difficult to imagine the workplace situations that Taylor was confronted with when he first began his analysis of industrial practices in the late 19th century. Workflows were not standardized or optimized. Machinery was usually custom built to order without working from a standard design, which meant replacement parts also had to be custom-built for that particular machine. Work got done when it got done rather than being planned to fit in with a schedule that would be coordinated with other workflows in the organization.

A prime example of the work of Frederick Taylor involves his analysis of working practices at the Bethlehem Steel Company.[4] The classic story, one which has ended up in many textbooks, involved his insights about the use of shovels on the factory floor of this particular steel company. By watching the workers, he noted that they were all using the same size shovel. Some workers would be using the shovel to move iron ore, and others might be using the same shovel to move ash from the furnaces. Those workers who were loading their shovels with iron ore quickly became exhausted, and their efficiency dropped. Those using the shovels for ash found that it was very easy to move the ash, but it took them a lot more trips to move it from one area of the factory to another.

Taylor's insight was that he could drastically improve efficiency by giving workers shovels designed for their particular task. He calculated that the optimum weight for a loaded shovel should be around 21 pounds, and so the weight and density of that material that was being moved (e.g., iron ore or ash) would determine the size of the shovelhead. For those men trying to move iron ore, using the new shovel designed for them meant they could carry on their task all day without becoming exhausted simply by having a smaller shovelhead and moving lighter loads more frequently. Conversely, those men whose job it was to move ash now found that their efficiency improved greatly because they had to make fewer trips, albeit with slightly heavier shovel loads than they had before. It was estimated that this simple change quadrupled the productivity of individual workers, and this huge leap in productivity significantly increased the profits of the Bethlehem Steel Company.

Frederick Taylor would go on to develop further his principles of scientific management, and it was later that Henry Ford, the legendary car manufacturer, began introducing similar practices of his own.[5]

It was Ford who truly understood the benefit of standardizing the design and manufacture of industrial equipment. Doing so meant that broken parts could easily be replaced by simply manufacturing the same part which would then fit neatly into place. Until then, industrial equipment had been crafted individually, which meant that if a specific part broke, the replacement also had to be crafted individually. With the standardization of the technical components of the sociotechnical system, efficiency improved greatly. This idea was then applied to his production line, where individual parts for his motor cars were mass-produced in a standardized fashion and then assembled quickly and cheaply by men on the line. The standardized industrial machines he had insisted on would then produce standardized parts which could fit together as planned no matter what machine they had come from. Again, standardization of processes and design led to a huge increase in efficiency.

In broader terms, Ford also realized that limiting customer choice made mass production easier as every car produced could be made to be the same. If he had wanted to give customers more choice about individual aspects of the design, that would have had to be factored into the production of materials and the production line itself. Ford felt that introducing such things would hurt efficiency.

While there can be no argument that Taylor and Ford created the principles of modern industrial engineering, there was one particular aspect of their drive toward efficiency, particularly on the part of Taylor, which may not have been so successful.

Taylor was obsessed with breaking down work processes into their component parts. He thought that by focusing efficiency measures on the component parts of the workflow process, he could improve efficiency overall. This involved stripping back any management involvement which he thought might be inefficient.

In principle, the idea of stripping down an organization to its lean components might seem a sensible one. This is similar to the idea that there is some sort of theoretic minimum of matter and energy within a system that is sufficient to allow it to function adequately. However, as we saw in the case of a mechanical watch, this idea may be true for technical systems, but it is not readily transferable to other classes of systems.

It is sometimes difficult to understand the value of management processes that are not directly involved with production but which instead seem to be all about conveying information upward and downward through the organizational system. We call these types of coordinating systems metasystems. For simple industrial processes, such as production lines, or a steelworks in Taylor's case, we might imagine that metasystems are less important. Material goes in at one end, is processed, and comes out in a different form at the other end. However, as sociotechnical systems become more complex, the role of metasystems in coordinating activity becomes all the more important. To be able to shift resources within a system to where they are needed, it is necessary to know what is actually happening in different parts of the system, and so these coordinating channels and metasystems become more significant. In modern sociotechnical systems, metasystems have an important function.

> Metasystems can be essential in coordinating activities across sociotechnical systems. They don't have to be seen as permanent features of a system but, conversely, they can't be viewed as unnecessary if they have been given a specific function to fulfill as long as that function continues to be valid.

2.5.4 Dynamic safety model

Jens Rasmussen, a noted researcher in industrial safety in the 20th century, developed a model which he thought could help explain how sociotechnical systems operate under pressure. He called it the Dynamic Safety Model.[6] Although this model may have its roots in the field of industrial safety, the principles are the same when we apply it to any sociotechnical system, even ones that are not safety critical. An outline of the Dynamic Safety Model is shown in Fig. 2.5.2:

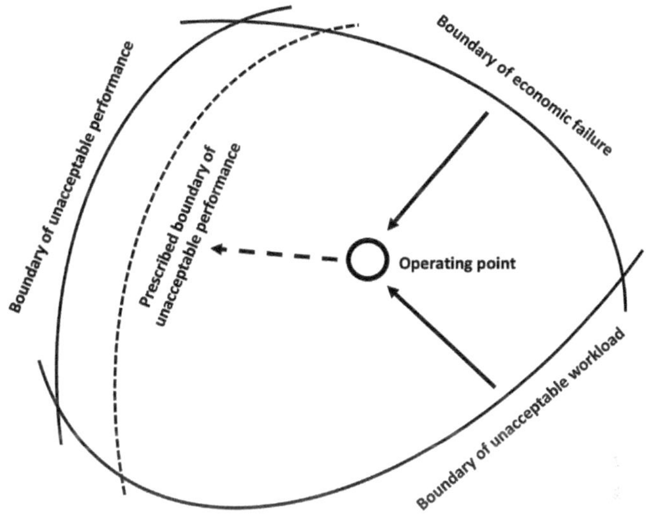

FIGURE 2.5.2 Dynamic safety model.

Rasmussen's view was that complex sociotechnical systems had an operating point that moved between three failure boundaries representing economic failure, unacceptable workload, and unacceptable performance. Pressures from within or from outside of the system will shift the operating point in any of these three directions. If the operating point ends up outside of any one of these boundaries, we can see how the system as a whole could fail.

Consider the pressures that might act from within the system; an overly strong union setup may shift the operating point toward the boundary of economic failure; strong pressure from management to improve organizational efficiency could move the operating point toward the unacceptable workload boundary or indeed toward the unacceptable performance boundary.

Another way the operating point could shift is when external pressures are exerted. Consider the effect of a toughening market condition on the operating point. The pressure here is being exerted on the outer part of the economic failure boundary and is pushing it inward. If the operating point doesn't shift, the boundary line itself will move toward the operating point as shown in Fig. 2.5.3:

As you can see from the diagram, the unacceptable performance boundary is, in fact, two boundaries: the prescribed boundary being closest to the operating point under normal conditions and the actual performance boundary being further away. What makes this configuration even more tricky is that we often do not know where the actual boundary of unacceptable performance is until we reach it and the system fails. In this case, the actual failure boundary is invisible.

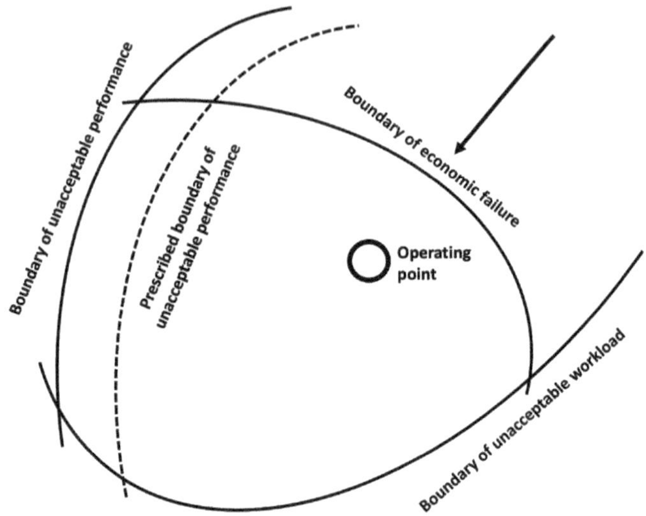

FIGURE 2.5.3 Dynamic safety model changes due to external disturbance.

The prescribed boundary is the limit set by management on what is acceptable. There is naturally a desire to have this boundary closer to the operating point than the actual boundary where system failure will occur. The space between these two boundaries gives us the error margin, which comes into play if unexpected conditions push the operating point too far toward the left.

Sociotechnical systems with operating points that frequently approach this margin can exhibit interesting dynamics. Rasmussen, together with another systems researcher, Richard Cook, suggests that the more frequently that the operating point of the system goes beyond the prescribed boundary toward the actual boundary of unacceptable performance, the more normalized this behavior becomes ("the normalization of deviance"—more on this in the case study on the Space Shuttle Challenger crash); the operating point gradually creeps further and further toward the actual boundary of unacceptable performance, thus increasing the probability that at some point a disturbance will drive the operating point beyond this boundary and lead to system failure.[7] This is a feature of procedural drift.

> Disturbances coming from outside or from within a system can push its operating point toward a failure boundary.

In the same paper, Rasmussen and Cook also described some characteristics of complex sociotechnical systems related to how far in any direction the operating point might move from its average position. By plotting these movements on the diagram, we can show the area throughout which we could expect the operating point to be under most conditions. This area might be tightly constrained or might cover a large area within these three boundaries. Systems operating in the Clear or Complicated domain would probably have an operating point that only moves within a small area within these boundaries, whereas a larger, more complex sociotechnical system could potentially have an operating point that drifts across the entire inner area enclosed by these boundaries. These two conditions are shown in Fig. 2.5.4:

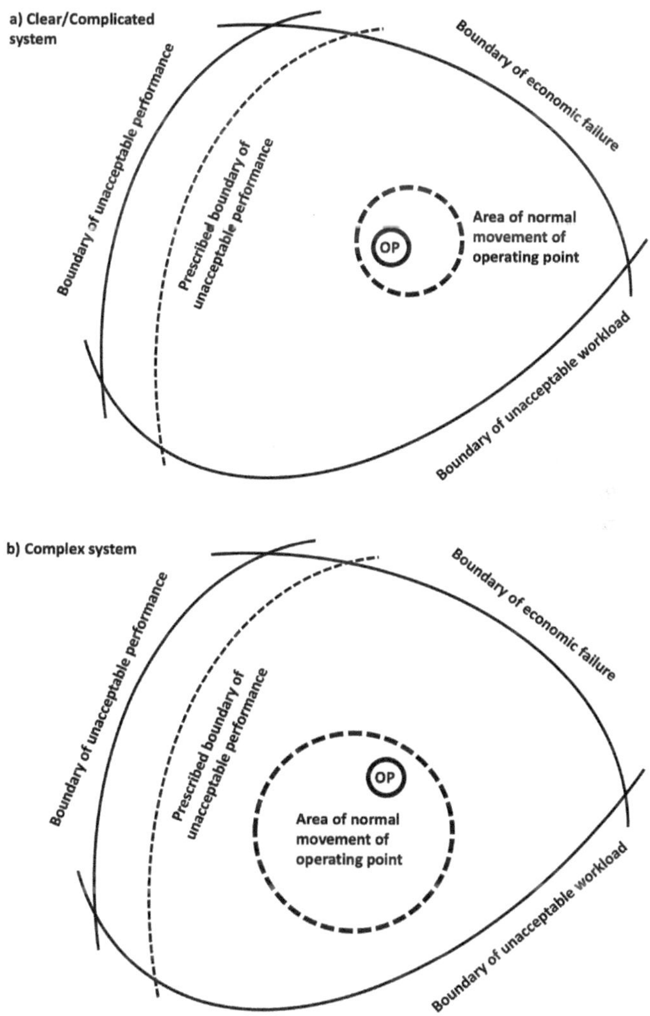

FIGURE 2.5.4 Area of movement of operating point (OP) in (a) a Clear/Complicated system; (b) a Complex system.

We can see that any sociotechnical system in any of the Cynefin framework domains will have an operating point that is somewhere within these boundaries. The extent of the movement of this operating point will be related to the Cynefin domain that the system is in. For example, as we saw in the diagrams, a system in the Clear domain is likely to have an operating point that is relatively static. Conversely, a system that is in the Complex domain is likely to have an operating point that moves quite widely within the extent of these boundaries. With this in mind, we can see that the role of management will be in defining the acceptable range of movement of the operating point; the tighter we make the acceptable range of movement, the more control we are likely able to exert over the system although this comes at a cost, particularly when it comes to the ability of the system to adapt.

> Under normal operating conditions, the operating point of a sociotechnical system will move slightly. The extent of this movement can give an indication as to what Cynefin domain the system is operating in.

We can also relate Charles Perrow's idea of coupling to the speed at which the operating point of the system might move. A disturbance in a highly coupled system may lead the system's operating point to move rapidly toward the boundary of unacceptable performance. This may happen so quickly that we cannot intervene and stop the operating point from exceeding this boundary and causing system failure.

Rasmussen's model is also useful because it gives us a clear visual representation of how pressure exerted by two different entities, for example, the management in one direction and the workforce in another, will inevitably drive the operating point in one direction; in this scenario, toward the boundary of unacceptable performance.

There are a few phenomena that can lead to movement of the operating point toward a failure boundary:

1. Trade-offs
2. Procedural drift
3. Domain shifts

2.5.4.1 Trade-offs

For as long as sociotechnical systems have existed, there has always been a balancing act going on between operational safety and operational efficiency (I include the concept of profitability within the term efficiency). To give an obvious example, let us go back to the sociotechnical system that is an airline. It would undoubtedly be safer if the crew of an aircraft performed only one flight a day. This would ensure that they are as well rested as possible for the following day. However, to do so would be highly inefficient. The crew might only work for 2 h out of a potential 8 h, in which case the airline would have to employ more crew to operate the rest of the flights. From the other perspective, it would be dangerous to have crews flying for every waking hour day after day. To find an acceptable balance, the aviation industry and the national aviation regulators established what are called flight time limitations. These are based on where the industry and the regulators have decided the balance between safety and efficiency lies; we can get the crew to do more than one flight a day, but we can't get them to do 10 flights a day. Adequate time must be built into their schedule for rest and recuperation. In sociotechnical systems that are not safety critical, the balance is between control over the activity of components in the system and their natural efficiency.

Another way that organizations try and achieve a balance between control and efficiency is through creating procedures. In most sociotechnical systems, having some sort of written (or otherwise known) procedure helps describe and prescribe certain aspects of activity. However, when organizations continue adding procedures to an already complex system, this can often result in the human operators being bogged down by the weight of the procedures they are supposed to follow. This is a paradoxical scenario where the system is being designed to be as predictable as possible, but when the humans in the system realize that to follow all these procedures would make the system too inefficient, they start taking shortcuts.

These shortcuts are known as efficiency/thoroughness trade-offs (ETTOs); the humans come up with a workaround which is a trade-off between being thorough (i.e., following the procedures exactly) and being efficient (getting the job done in the time allowed).[8] This can lead to a scenario in complex sociotechnical systems where the actual operating point of the system is very different from where management might expect it to be based on the procedures they have laid out. This is the difference between work-as-imagined (WAI) and work-as-done (WAD). It should be noted that the gap between these two concepts (and, indeed, the gap between the imagined and actual operating points) is not down to incompetence or a desire to break the rules on the part of the human operators. It is simply a way to balance what is expected of them, however unreasonable those expectations may be, and what is actually required to get the job done. While management might assume that always following the rules is the best way to keep the operating point within an acceptable margin, in fact, without a clear understanding of work-as-done (the ETTOs being made), the operating point might be nowhere near where they think it is.

Erik Hollnagel, a professor of industrial safety who first developed this idea of efficiency/thoroughness trade-offs, gives an excellent example of how ETTOs are important to keep sociotechnical systems working.[9] He asks us to consider a scenario where the employees of an organization wish to undertake industrial action but do not wish to be fired for doing so. A common way of doing this is to carry out what is called a work-to-rule strike. In this sort of industrial action, the human operators in the system simply say that they will follow every rule and procedure exactly. The result is that efficiency/thoroughness trade-offs are no longer made, and the system grinds to a halt.

In many cases, the operating point of the work-as-done (i.e., the true operating point) may be further away from the boundary of unacceptable workload than expected while still keeping away from the boundary of unacceptable performance. An agile organization may recognize that there is a difference between work-as-imagined and work-as-done and take the opportunity to investigate the trade-offs that the human operators have developed and subsequently adjust the written procedures to account for them. Less evolved companies may use this difference between work-as-imagined and work-as-done as a stick to beat employees with when things go wrong.

Fortunately, in industrial safety, we are moving away from the idea that following the rules at all times will always keep us safe, and later we will look at error and error management in further detail. This idea also extends to systems that aren't safety critical. Suffice it to say, efficiency/thoroughness trade-offs are what keep many complex sociotechnical systems functioning because it is normally the humans who have the flexibility to adjust their operation as conditions require.

> Efficiency/thoroughness trade-offs are often essential for keeping a sociotechnical system operational.

2.5.4.2 Procedural drift

Procedural drift is related to trade-offs in so far as it involves an increase in the gap between work-as-imagined and work-as-done. Many organizations carry a gap between these two conditions, and, in many cases, it makes little difference. In fact, many systems depend on this gap as we have seen from the example of the work-to-rule strike action. The real risk though is when this gap between work-as-imagined and work-as-done is slowly increasing, perhaps without the realization of the human operators or management.[10]

An increasing divergence in work-as-done away from work-as-imagined can be defined as procedural drift. Unlike trade-offs, which are sometimes a necessary part of the operation of a system where the time available to complete tasks is not sufficient to allow operators to follow every procedure, procedural drift is a more gradual, normally unintentional divergence from work-as-imagined. It may not be that this divergence is helping the system's efficiency, but rather it represents how once a small deviation from procedures occurs, it is easier then to imagine a larger deviation. Suddenly, the now divergent operating point is the new normal, and, in this way, the operating point of the system gradually drifts toward a failure boundary. This has sometimes been called the normalization of deviance.[11]

The phenomenon of procedural drift has much in common with the phenomenon of entropy. Without the energy-dependent recompartmentalization of set behaviors, the system gradually becomes increasingly disordered in a potentially random way, and this drift of the operating point may lead to it crossing a failure boundary. We can contrast this with the phenomenon of efficiency/thoroughness trade-offs, which are deliberate choices that operators make to deviate from procedures because the procedures aren't workable in the time available. This is an active process, whereas procedural drift is passive.

> Procedural drift is a passive process where the gap between work-as-imagined and work-as-done gradually increases, and this can push the operating point toward a failure boundary.

2.5.4.3 Domain shift

We introduced this concept when we covered the Cynefin framework. To briefly recap in the context of how it affects the operating point, a domain shift will potentially increase the size of the area within which the operating system moves when it is considered to be operating normally. The problem lies when this domain shift occurs without being recognized. The normally small area that represented the normal movements of a system in the Complicated domain may have suddenly gotten bigger as a domain shift has occurred into the Complex domain. Suppose the initial operating point had been near a failure boundary. In that case, it may be that this increase in size means that part of the area within which the operating point moves is now beyond a failure boundary, meaning that failure could occur unexpectedly or that another domain shift could occur into the Chaotic domain. This is shown in the Fig. 2.5.5:

> Domain shifts can increase the area across which the operating point moves to the extent that it can make it more likely to cross a failure boundary.

2.5.5 The role of humans in sociotechnical systems

To finish off this chapter, we should for a moment consider the role of humans in sociotechnical systems. Opinion has changed over the years about

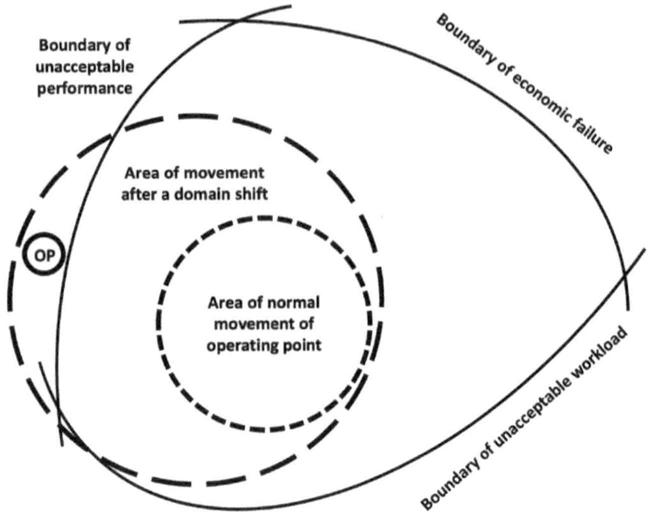

FIGURE 2.5.5 The effect of domain shift on the movement area of the operating point (OP).

the roles and responsibilities of humans in these systems. The archaic view was that it was possible to design a system that would be both safe and efficient. Rules could be established, staff could be trained, and processes streamlined. This finely tuned system would then run without problems. This type of thinking led to the conclusion that if something negative did happen to the system, there was no way this could be down to the way it was designed. The only thing that could have happened is that there was a human failure. The logical next step would be to get rid of that human from the system.

Based on the insights of Charles Perrow and subsequent research into the dynamics of sociotechnical systems, we are moving away from this view that the human is the weak point in the system. By virtue of their complexity, sociotechnical systems are not inherently efficient or failure-proof, and we cannot engineer anything into the system to guarantee this.[12] Much like many of the phenomena that we considered in the physical world, we are left with uncertainty and probabilities.

> Sociotechnical systems, particularly complex ones, cannot be designed to be inherently reliable all of the time. We can only manage probabilities of success and failure.

Another perspective that has changed over the years is more specific to the humans in the system. Although we recognize that there is a natural variability in human performance from day-to-day, maybe hour-to-hour, we can also see that this flexibility in performance can also be a strength when dealing with disturbances. The mechanics behind this will be covered later on in the book.

> The variability of human performance can be a source of disturbance within a system. However, it is this same variability when deployed as flexibility, which can also act as a buffer against disturbance.

2.5.5.1 Joint cognitive systems

One final subtype of sociotechnical system which is worth looking at specifically involves those systems that have both human control and computer control systems regulating their operation. These are distinct from sociotechnical systems that are highly automated and where the human can see what the computer is doing but is unlikely to need to intervene. In systems that require both human and computer control, there can be a risk of mismatch between these two classes of control system. Such mismatches can be highly problematic.

When designing these systems, they need to be built as joint cognitive systems. A joint cognitive system recognizes that there are inherent differences in how quickly humans and computers analyze problems and make decisions.[13] It also recognizes the limitations of computers to make decisions on anything other than the rules with which they have been programmed. To use the analogy of Jens Rasmussen's levels of performance, computers cannot apply knowledge-based problem-solving in novel, dynamic situations (although artificial intelligence systems may overcome this limitation).

A joint cognitive system is designed to slow the pace of computer control (unless absolutely necessary) to allow the human operator to be able to update their mental model of the situation at every step. To ensure this, some systems force the human to take particular steps (e.g., closing valves or switching on mechanical systems) rather than having the computer do it automatically. This ensures that the human operator is keeping up with the changes that are happening within the system.

In certain sociotechnical systems that require human intervention, the failure to design control programs along these lines can lead to significant problems as we will see in the case study of the Three Mile Island nuclear disaster.

> Systems that require both human and automated control should have their control systems designed as joint cognitive systems.

Chapter 2.6

Consilient dynamics across scales

Bottomless wonders spring from simple rules which are repeated without end
Benoit Mandelbrot, mathematician specializing in fractal geometry

Chapter outline

2.6.1 Introduction

Sometimes it is useful to see past the arbitrary classifications we put on ways of looking at the world. We know that, to a certain extent, we need to simplify to understand and that the choices we make in how we simplify affect how we see what we are looking at. A biologist will simplify and model a phenomenon differently than a sociologist might do. An artist will likely see something different in an astronomical event than an astrophysicist would. The goggles we wear can help us orient ourselves in our dynamic, novel universe by filtering out aspects of what we are seeing, but we need to remember that the filter settings for those goggles are based on our backgrounds and knowledge and are unique to us. An English botanist notices that tiny pollen seeds suspended in water moved in an irregular swarming pattern and thinks no more of it. Einstein saw this same phenomenon and deduced that the pollen had to be being bombarded by individual water molecules. His theory on this phenomenon established the evidence that atoms and molecules actually exist.[1] This example would have been even better if the Roman poet Lucretius, who described the motion of dust in the air, had tried to use the theory of atoms to describe their movement.[2] In that case, it was the right explanation in the wrong setting. Air currents were moving the dust.

From physical to chemical, to biological, to social, to technical, and finally to sociotechnical systems, we can see that one obvious commonality exists between them: they are all systems. Despite our necessary drive toward specialization, one could imagine superintelligent beings on other planets who

Designing and Managing Complex Systems. https://doi.org/10.1016/B978-0-323-91609-7.00007-X

do not need to specialize. As knowledge on their planet grows, so their brains grow to accommodate it all. There are no physicists or artists on their world. Instead, everyone knows everything. Would such beings describe the phenomena that we see in the physical and social spheres as being truly distinct? Instead, might they not see an underlying dynamic that exists and manifests itself in slightly different ways when viewed from different scales?

Of course, we have no unifying evidence that these underlying dynamics exist. We only have anecdotal evidence. Fortunately, we are not concerned with evidence. We are concerned with utility. The job of creating a grand unified theory for how matter, energy, and information interact across all scales across all of reality is a job for another day! My sense is that even these anecdotal connections can be useful and that there may be more connections still to be discovered.

To that end, as we come to the end of this part of the book, we can look back on how much material we have covered. We have taken a journey from the birth of the universe right up to the present day and the modern, complex sociotechnical systems within which we work. Many of these phenomena are directly applicable to Part 3 of this book. Others, I think, may have an impact. It could also be that the reader will identify commonalities that I have failed to notice.

Clearly, I have not covered every phenomenon that affects all systems across all scales. For the purposes of this book, I could only trace a narrow path through all the events that have spread out over time as the universe has unfolded itself since the Big Bang. I would say it is a certainty that other fundamental dynamics exist across scales that this book hasn't touched on. However, what we have touched on should be enough to act as a launching pad for what follows.

2.6.2 Summary of concepts covered

Before moving on, let's have a look at some of the concepts we have introduced in this part of the book:

1. The limitations of models – we necessarily must simplify, but we need to accept that if we create our models on the wrong scale, we are going to have too much or too little information to be useful to our analysis.
2. Chosen, forced, and entrenched simplification – the simplification imperative is strong in human beings as it is sometimes the only way that we can filter out certain aspects of our dynamic world to allow us to focus on whatever it is that is concerning us. However, it can drive us toward models and explanations that don't have enough detail.
3. Necessary complexity – we cannot reduce every phenomenon of interest to its component parts and think that this will better help us understand it. Sometimes it is the complex nature of the system's interactions as it is operating that is at the heart of its fundamental dynamics. We sometimes must embrace the messiness of this complexity to be able to manage it.

4. Cybernetics – there are consilient dynamics in the world of biological control that are transferable into the world of sociotechnical control.
5. Systems thinking – we can view the world as interlocking systems, no matter which sphere of reality they seem to exist in. If we start thinking in these terms, we can better see how phenomena in different areas are actually related.
6. Language of systems (open, closed, etc.) – if we take a systems-thinking perspective, systems have certain characteristics that are fundamental to understanding them. This is the *lingua franca* of systems.
7. Recursion within systems and subsystems – as we zoom closer into looking at aspects of systems and subsystems, we find the same patterns emerging as features of the system as a whole seem to be repeated at smaller scales within its subsystems.
8. Metasystems – these act between subsystems to coordinate activity. A reductionist approach to improving efficiency in systems might try and strip out these "accessory" systems as it might not seem obvious how they relate to the actual work of the system. However, these coordinating metasystems are important in balancing day-to-day activity in the face of changes to the overall running of the system and, more significantly, can be vital in coordinating activity if the system encounters disturbances.
9. The Law of Requisite Variety - the more we expect our systems to do and the more resilient we expect them to be in the face of disturbances exerted by the dynamic environment in which they function, the more complex the system must become.
10. Multiple goal-seeking – different components/subsystems within a system may have different goals.
11. Equifinality – biological and social systems operate differently from mechanical ones because system endpoints can be achieved from various starting conditions, i.e., two systems exhibiting equifinality can reach the same endpoint from different starting conditions.
12. Tractability – how understandable a system is in terms of the number of connections between different components and subsystems.
13. Coupling – how tight these connections between components and subsystems are. When a disturbance happens in one part of the system, how quickly does that disturbance spread?
14. The observer effect – the perceived nature of what is being observed will always be colored by the knowledge state of the observer.
15. The Cynefin Framework – suggests that systems operate within a particular domain: Clear, Complicated, Complex, or Chaotic. This will affect our choices in deciding how to manage these systems.
16. Domain shift – this can occur when a system shifts from one Cynefin domain to another, usually because of some sort of disturbance. We need to ensure that our management style changes with it.

17. Effective complexity – the more we want our system to do, the more complex it needs to become. Adding humans to a system is a good way of increasing its complexity because of the flexibility of human performance.
18. Different classes of system – despite recognizing that our perspective on classes of system is affected by how we have divided up different areas of human knowledge, we are defining seven classes of system: physical, chemical, biological, social, informational, technical, and sociotechnical.
19. Energy state-dependent spontaneous self-organization – spontaneous self-organization is more likely to happen if the energy within the system is at an appropriate level.
20. Quantum mechanical uncertainty and the nonmechanistic view of the universe – quantum mechanics tells us that we can't view the universe as a clockwork, closed system with every path of every atom being predictable based on their first positions. The fundamental physics of matter is built on probability and uncertainty.
21. Essential, indivisible systems – at the core of matter, there is a system of indivisible components. This system can only be considered as a whole. Again, it suggests that a reductionist approach fails to get us to the point where we can explain reality by analyzing singular units of matter.
22. Autocatalysis – the process of spontaneous self-organization and replication in chemical and early biological systems.
23. Natural selection – the evolution of systems based on their fitness to function in their environment.
24. Minimum gene sets and viability – the minimum set of genes that an organism needs to possess to create the proteins that allow it to be viable in its environment. Even with monocellular organisms, there are some of these genes/proteins whose function remains unknown, but without which, the organism won't survive.
25. Entropy – how organized energy in the universe is. We can only use energy to do any sort of work if we can find an environment where the amount of energy in one compartment is different from the amount of energy in another. By allowing the boundaries that encapsulate this energy to become permeable, the natural process of entropy will drive the two energy zones to try and equilibrate. As energy flows between the two to try and find a balance, we can use this movement to allow us to carry out some sort of work. This work could be a chemical reaction or mechanical movement.
26. Total entropy and local entropy – we can decrease local entropy by creating compartments of different energy levels that we can later exploit to do work. However, total entropy in the universe is always increasing.
27. Constraints – to exploit energy to do work, we need to constrain how that energy is directed. Placing an explosive charge under a cannonball will certainly move the cannonball. However, placing the explosive charge at the bottom of a cannon and putting the cannonball on top of that charge allows us to constrain the release of energy and direct the cannonball in the direction that we wish it to go.

28. Out-of-the-silo perspective on complexity increasing throughout history – we can strip away human distinctions of physics, biology, and chemistry and consider the upswelling of complexity in the universe in terms of systems and how they use energy.
29. Energy-dependent boundary building and maintenance – any compartmentalization of matter or energy requires energy. It also requires energy to maintain these boundaries.
30. Informational compartmentalization within human societies – no matter how information is recorded and stored, either physically or mentally, the compartmentalization requires energy, and it also takes energy to maintain this compartmentalization.
31. Theoretic minima in technical systems – depending on the system's requirements, there is a theoretic minimum amount of energy and matter required to create that technical system.
32. Divisions in the human nervous system (central, autonomic, and peripheral) – although interconnected, the roles of these three systems are distinct, and the central and autonomic nervous systems, in particular, will be relevant to Part 3.
33. Functional scales of activity throughout the nervous system – organizational scales within the nervous system follow similar principles to organizational scales in sociotechnical systems, including recursive subsystems being found at different scales.
34. Neurons as complex adaptive systems with identity – the flexibility of neurons to strengthen, weaken, or otherwise change the nature of their connection to other neurons in ways that are incompletely understood means that they are complex systems in themselves. The unique internal physiology and connectivity within the human connectome give individual neurons an identity, meaning that they cannot simply be replaced with a different neuron and then behave exactly the same way.
35. Plasticity – the brain, particularly in younger people, can respond to damage by redistributing functions to other, undamaged brain areas.
36. Autonomic control and communication – most essential physiological functions are controlled automatically by the autonomic nervous system. The autonomic nervous system can change the operating parameters of the systems over which it exerts control in response to external or internal disturbances. There are dedicated communication channels that allow this to happen.
37. Complex structure leading to function – the ACT-R model has taken us some way in explaining how and where various brain centers coordinate to produce complex behavior. Seemingly complex behavior derives from interactions between highly coupled, interrelated neuronal networks.
38. The anatomy of decision-making – ACT-R also shows how decision-making can be explained by looking at the arrangement and interaction of several distinct brain areas.

39. The two modes of decision-making and the impact of workload – humans have two modes of decision making: a quick, instinctive mode, and a slow, rational mode. In high workload situations, we are more likely to rely on the first mode despite the higher risk of it giving us incorrect responses.
40. Algorithmic decision-making – in many diverse settings, decision-making using systematic algorithms proves more reliable than human decision-making due to the innate biases that affect human decision-making.
41. Levels of performance – human beings' responses can be divided into skill-based, rule-based, and knowledge-based performance, depending on how much mental effort is required.
42. Homeostasis and allostasis – the autonomic nervous system controls certain physiological functions to keep them within certain parameters. However, if the homeostatic mechanisms are constantly having to do this because of repeated disturbances, the body eventually employs allostatic processes which adapt the anatomy and physiology of the body to account for these new and persisting disturbances.
43. Negative feedback – loop-like control of a parameter that engages processes that that will return that parameter to within set constraints before disengaging. Occurs across many different system types.
44. Positive feedback – loop-like amplification of a parameter by constantly increasing its value until resources are exhausted, or some other endpoint is reached. Positive feedback is more likely to be associated with disturbances, particularly in biological systems.
45. Signaling through the reticular activating system – the reticular activating system in the brain has several functions, one of which is to act as a general alerting system. Even if we are asleep, the reticular activating system is monitoring the sensory inputs, which are still flowing to the brain, and if it detects something which it thinks requires our response, it will initiate the process of waking us up.
46. Reflex responses – some stimuli-response reactions are not processed in the brain. Instead, there is a rapid reflex loop that just goes through the spinal cord.
47. The nature of sociotechnical systems – sociotechnical systems are those designed to carry out a productive function, normally related to work or industry.
48. Different component types in sociotechnical systems – sociotechnical systems are made up of human, machine, and procedural (including software, laws, etc.) components.
49. Shifting operating point – the operating point of a sociotechnical system is not static with respect to the boundaries past which the system will fail. The operating point can be expected to move a little by itself in the normal course of operating, but disturbances from both inside and outside of the system can move it toward a failure boundary.

50. Work-as-imagined versus work-as-done – work-as-imagined is what a manager would think is happening in an organization based on the procedures that have been established. Work-as-done is what is actually happening. This can be represented by the real and imagined operating points.
51. Failure boundaries – there are three failure boundaries that could affect a sociotechnical system: boundaries of economic failure, unacceptable workload, or unacceptable performance.
52. Prescribed, actual, and invisible boundaries – some boundaries are established by procedures, some boundaries are fixed by the nature of the system, and some of these boundaries are invisible.
53. Normalization of deviance – an operating point that is persistently near or past a prescribed boundary is likely to stay there as this new position becomes the norm.
54. Procedural drift – the operating point of a system may gradually drift toward a failure boundary as people begin to take unnecessary shortcuts or forget parts of the procedures.
55. Efficiency/thoroughness trade-offs – these could be viewed as necessary shortcuts which happen to keep the system operating. It is not always possible to do the work exactly as procedures suggest as there may not be enough time available.
56. Control systems relying on human and automated inputs need to be designed as joint cognitive systems.

Many of these concepts apply across many different scales of many different classes of systems. For example, we can see the impact of local decreases in entropy whenever we see complexity increasing at any scale in any area; by necessity, work must be done to create order. The same thermodynamic law that says total entropy will always increase can be seen to have an effect when compartmentalized information begins to decay when not reinforced (forgetting a language or how to carry out a rarely performed task at work if that knowledge hasn't been refreshed frequently enough).

We can derive most of these concepts from observation of the world, whether it be the natural world or the world we have created for ourselves in our societies. In the next part of this book, we will see how we can apply these concepts to the goal of creating better systems and managing them more successfully.

Part III

Creating and managing systems

Chapter 3.1

Introduction to part 3

Chapter outline

Knowing is not enough; we must apply.

Johann Wolfgang von Goethe, polymath in $18^{th}/19^{th}$ century German

3.1.1 Introduction

In this part of the book, we will start to draw together the threads of the concepts that we looked at in Parts 1 and 2 and combine them into something that we can use to design and manage complex systems. It has been somewhat of a challenge trying to find a logical way of presenting these ideas, but when in doubt, the natural world probably has an example of something similar that works. We can understand organisms by considering their anatomy from the bones out toward the organs, their physiology from how the organs work to how they are controlled, and finally, to the overall dynamics of the organism as a whole. In the same way, we can look at systems from their bones outward. We will look at the structure of viable systems, their functions and sub-functions, what they need to carry out these functions, and finally, how they behave when they are operating and dealing with the complexity of the real world.

3.1.2 Definitions

There are a few definitions we need to clarify before moving on. We have covered most of these in previous sections of the book, but having a solid understanding of these definitions now will make this part of the book much clearer:

- Structure – the actual arrangement of components within the system. These could be people, machines, or procedures/software/rules.

Designing and Managing Complex Systems. https://doi.org/10.1016/B978-0-323-91609-7.00008-1

- Function – what we want a system, subsystem, or component to do.
- Capability – the ability of the system, subsystem, or component to fulfill its function.
- Dynamics – how a system, subsystem, or component deploys its capability to fulfill its function. This includes its adaptive capacity and overall resilience.

A system is designed to fulfill a function. If we have designed it successfully, the subsystems and components will have the capability to fulfill their functions within the system as a whole, and in so doing, the system can fulfill its function. Under normal conditions, the flow of matter, energy, and information in the system is dynamic. By understanding how this works to keep the system operational, we can delineate the system's dynamics.

These dynamics may well change if the system must react to an unexpected disturbance, and so the dynamics of a system are not fixed.

3.1.3 Dynamics

Although "dynamics" may appear similar to "function," there is a subtle distinction between the concepts. A body will function well in a regulated environment that permits the supply of oxygen, food, water, and heat. Things become more complicated when the body is in an environment that isn't regulated. While homeostasis governs the stability of specific physiological functions (blood sugar, temperature, pH, etc.), allostasis (introduced in Part 2) is a more nuanced process that involves changing a wide range of physiological parameters over a longer period so that the organism can adapt to persistent external disturbances. For example, humans under prolonged stress will alter various hormone pathways and change the way certain genes are expressed to alter their physiology to cope with this stress. In this way, allostasis reconfigures the structure and function of some of the components and, in so doing, changes the dynamics. Whereas previously, the body had been able to remain viable using dynamics suited to its preexisting structure, the interplay of allostatic dynamic change leading to a rearrangement of structure and alteration of function means that the reconfigured system will have dynamics different from the original system.

3.1.4 Building up our understanding of systems

To be able to put this example into the context of systems engineering, here is a summary of the kind of narrative that we will be developing over the following chapters. It may seem counterintuitive to present the "big picture" in the introduction, but this brief example that includes many of the concepts that we are going to explore in more detail should help you to build up a more complete understanding as you work through this part of the book:

- Let us suppose that the purpose of the sociotechnical system in this example, like most sociotechnical systems, is to complete an industrial function while still remaining viable. For example, for a car manufacturer, it would be the production and profitable sale of vehicles.
- Sociotechnical systems have a recursive structure. The system contains subsystems, which also contain subsystems and so on.
- Ultimately, the final "level" of subsystems can be seen to contain components.
- Components can be humans, machines, or procedures/software/laws.
- Components have assigned functions and need the capability to carry out these functions.
- A component is the simplest part of a system that can be assigned a single function by the designers and/or operators of the system.
- At the local level, components will need to adapt their activity to continue to be able to fulfill their assigned functions based on the demands of the subsystem of which they are a part.
- The degree to which the component's assigned functions allow for flexibility (later defined as adaptive capacity) will determine how adaptive they can be to respond to the demands of the subsystem of which they are a part.
- Normal activity within a stable sociotechnical system under no external pressure should be possible using only local adaptive capacity (local to the component or its subsystem) to smooth out the operation.
- In the case of a nonstable sociotechnical system (one that is expanding or contracting, or a sociotechnical system that has experienced internal failures or must respond to external pressures), the local adaptive capacity at the level of components is unlikely to be sufficient to allow the system to continue to complete its industrial function and remain viable.
- The broader, system-wide effects of these factors are the system's dynamics. Changes in dynamics are not always readily apparent and may happen slowly over time. Low morale in a workforce may gradually cause procedural drift as employees are less engaged in the operation. Procedural drift may eventually lead to a component failure. Insufficient adaptive capacity may mean that the effects of the failure cannot be buffered at the local level. Resources from other subsystems may be needed. This redistribution of resources may leave other system areas vulnerable and result in wider failures.
- In Part 2, we saw that homeostasis involves local changes to try and maintain stability, and allostasis involves systemic changes to try and maintain viability. In sociotechnical systems, adaptive capacity at the local level tries to maintain stability. If there are wider effects, if principles of resilience are built into the system, these will allow the system to monitor changes in the dynamics of the organization, respond in order to deal with these changes, and ultimately, learn from this process to be able to anticipate better in the future.

What follows in the next chapters gives more detailed guidance about each step of this introduction.

Chapter 3.2

Structure and function

Instead of trying to specify a system in full detail, specify it only somewhat. You can then ride on the system's dynamics in the direction you want to go.

Stafford Beer, cybernetician

Chapter outline

3.2.1 Designing structure and assigning function

How would you describe the structure of the organization within which you work? Is it the organizational chart that shows who is responsible to whom? Is it the operations manual that describes how things should be done and who should do them? Is the structure less rigid than either of those to the extent that lines of authority and responsibility are not clearly defined?

For most of us, the chances are that we work in an organization that is organized hierarchically and is often represented as an organizational chart, as shown in Fig. 3.2.1:

We seem to think that this represents the organization's structure, but, as many have pointed out, it is more useful as a blame-finding tool. When A notices that something has gone wrong, she will ask B to investigate. B might

Designing and Managing Complex Systems. https://doi.org/10.1016/B978-0-323-91609-7.00009-3

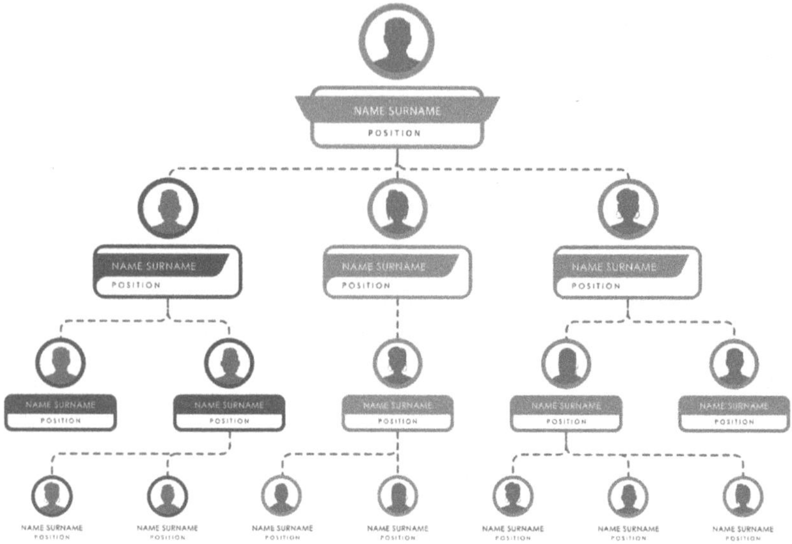

FIGURE 3.2.1 Example of an organizational chart. *Image copyright Hong Choo Liang (www.shutterstock.com).*

track down C, one of the heads of department. C then takes a closer look at her subordinates, D, E, and F, and decides that it is the fault of F simply because she happened to be out of the office when C was looking for a scapegoat.

Organizational charts are great for doing things like that, but they often do not represent the organization's structure in a meaningful and, more importantly, useful way.

We also need to consider how systems come together. In Part 2, we looked at the idea of spontaneous self-organization, i.e., that components can come together (autocatalyze themselves) to form a system if the conditions are right. The sociotechnical systems within which we work are usually a combination of design and evolution. A system designer can anticipate the components that are likely to be required for the system to be viable, but it may also be the case that once the system begins to operate, it undergoes a process of evolution and spontaneous self-organization as the system refines itself when it starts interacting with its operating environment. What follows in this chapter on structure and function is an exploration of the "minimum gene set" required for a potentially viable system. Depending on the type of system that is being constructed, certain components and subsystems will likely be more important than others, and once the system is running, certain additional subsystems may form while others decay according to the demands of the operating environment.

We should accept that as circumstances change, so systems will change, possibly without the intention, or direct intervention of the people managing them. Stafford Beer's quote at the start of this chapter is very true: designing structure and function is just the start of the process. However, if we can get the right minimum gene set working from the outset, the subsequent evolution and refinement of the system should be a much easier process.

3.2.2 The Viable System Model

When Stafford Beer was developing his Viable System Model (VSM – see Part 1), his ideas came after many years of analyzing management and operational structures. He identified certain interconnected features that were present in some, though not all, of the organizations he looked at and whose presence or absence had an impact on the success of that organization.

Using the human nervous system as a model, Beer developed the VSM as an exemplar of a viable organizational system. This chapter uses the word "structure" to describe the physical arrangement of components (humans, machines, and their associated procedures/software). The VSM was also a way of mapping out a structure that would make it more likely that the natural dynamics that such a structure would engender would ensure that the components and subsystems had the capability they needed to fulfill their functions and, hence, the system as a whole would be able to fulfill its function. The original VSM was based around five systems, shown in Fig. 3.2.2:

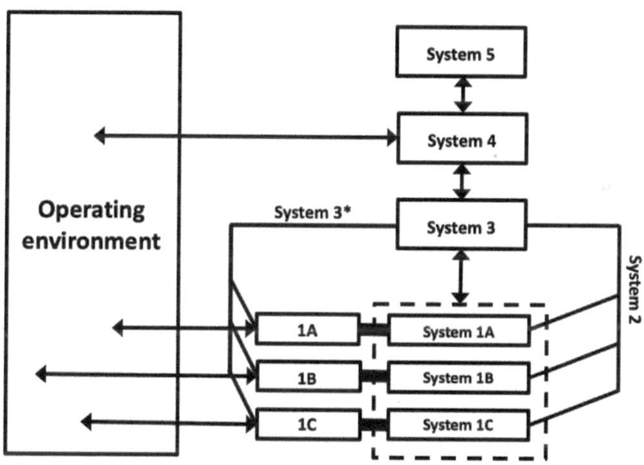

FIGURE 3.2.2 The Viable System Model.

The systems are as follows:

- System 1 – A large organization will have many System 1 entities which represent individual functions. In an airline, departments such as fleet management, engineering, training, flight operations would all be System 1 entities. It is important to note that System 1 entities are in themselves viable systems and so will have their own Systems 2–5. In this way, the VSM is often called a recursive model as System 1 entities may have a similar structure to that of the organization as a whole in terms of policymaking, adaptive capacity, planning/control, and information channels. This is shown in the diagram by showing each System 1 entity as having two parts: the operational part and the regulatory part.

- System 2 – The information channels that entities at System 1 level can use to organize themselves, e.g., allocating shared resources, communicating needs and requirements between departments, etc.
- System 3 – Can be defined as operations planning and control. Establishes rules and distributes resources to System 1. (System 3* represents the audit functions of System 3, i.e., how information regarding what is actually being done at System 1 level is communicated to Systems 4 and 5).
- System 4 – Uses environmental data to predict how the system must adapt to remain viable.
- System 5 – Responsible for system identity and high-level policy.

Beer was also at pains to point out that the VSM was not meant to be a hierarchical structure in terms of management processes. The idea was that System 1 entities would organize themselves to carry out their function while the rest of the systems were only there to ensure that System 1 entities had the resources necessary to do this and occasionally intervene if necessary. Although the arrangement of the VSM appears to be hierarchical, this was simply a way of showing how it was modeled on the arrangement of the human nervous system. I have seen arrangements of the VSM in which the recursive structure of the system was the key feature of the diagram, and this led to a recursive pattern leading out from a central point, a very different visual arrangement but one that would be functionally the same as Beer's original diagram.

Beer reasoned that the VSM was not just a tool for organizations. During his time working with the Chilean government, it looked likely that he would get the opportunity to apply this model to the management of the economy of an entire country until the military coup ended those plans.

The VSM has been a highly successful aspect of scientific management practices. Beer himself was a cybernetician and was part of the core group of scientists and researchers who grew the nascent field of cybernetics into an industry in itself. Management cybernetics was itself a branch of the cybernetics movement, a field that grew from the idea that there were similar (consilient) phenomena related to control and communication happening in all different classes of systems. It was by realizing that the structural components that make a biological system viable could be adapted to make an organization viable that Beer made his next logical step. He concluded that the human nervous system was the key to viability in a highly dynamic and sometimes hostile world, and so by adopting its general structure, organizations could exhibit the same viability in a dynamic commercial world.

The VSM was hugely influential when Stafford Beer first introduced it and has formed a cornerstone of management cybernetics. Beer had based his design for the model on what was known about the structure and function of the human nervous system at the time he was developing it (the late 1960s). Since the original VSM was published in 1971, research in neuroscience and psychology has explained much more about how the nervous system functions, and we have covered several of these new findings in Part 2. As you may have figured out by now, this book has been driving toward presenting an updated version of Stafford Beer's original VSM. There was a lot more to the VSM than the five systems as described here, but as our updated model is going to include many of the same features that made the original model so successful, it will be easier to explain these features in the context of the revised model.

3.2.3 The Revised Viable System Model (rVSM)

The original VSM had a couple of drawbacks. One was the language that Beer employed to describe it. Coming as it did off the back of a fruitful and revelatory academic life exploring, and at the same time, building the new field of cybernetics, the language sometimes used to describe the various components, subsystems, and dynamics of the VSM could be challenging to someone not familiar with the field. Now, even more than then, the language of attenuators, variety, algedonics, amplifiers, efferents, afferents, and so on is challenging to understand. As such, the rVSM will use some different language to describe some of these features. Additionally, to avoid drowning in "systems," the rVSM uses some alternative language for the five systems that Beer originally described.

By presenting the revised VSM, I hope to be able to introduce you to the phenomenal achievement of Stafford Beer in a way that will be easily accessible to you and which benefits from the findings of contemporary research into biological, technical, and social systems. The arrangement of the various components remains broadly the same, but in describing the functions of these components, I have attempted to update them with the modern research findings that we covered in Part 2. I'm also conscious of the quote from Stafford Beer at the start of this chapter. It may look like the system we build up is overly complicated when compared to our nice, simple organizational chart. However, from an enterprise architecture perspective, the plan of the VSM (and, hopefully, the rVSM) represents the minimum gene set for viability. Many elaborations could be added on, but the structures described in the following pages are the base template on which the system is founded. As Beer says and as we saw in Part 2, once the system is running and exchanging matter, energy, and information with its environment, its dynamics may cause the structure to change and evolve. Like an autocatalytic set, once it starts, with the basic structure in place, it will go on to build and refine itself.

It is also worth noting that this chapter is based upon the idea of building a system from the ground upwards. If we accept that structure is outlined in the minimum gene set required for viability, you would expect to find all these subsystems and functions distributed throughout any successful organization. They may not be represented on the organizational chart in the same way as they are arranged here (or they may not be represented at all), but someone or something in the system must be performing these functions. Many consulting firms use Stafford Beer's original VSM as a starting point for trying to identify why their client's companies are struggling. By trying to ascertain what parts of the company fulfilled the functions of systems 1–5, they would often find that some part of this minimum gene set of components was missing or working poorly.

The rVSM term for System 1 is SYS1, pronounced "siss 1". As you will see from the diagrams, there is likely to be more than a single SYS1 in an organization, so all the SYS1 entities together are known as the Operating Body, and SYS3, 4, and 5 are known together as the Regulatory Body. SYS2 connects them. The general arrangement of the rVSM is shown in Fig. 3.2.3, and the recursive nature of the SYS1 entities is shown compared with the general rVSM layout in Fig. 3.2.4:

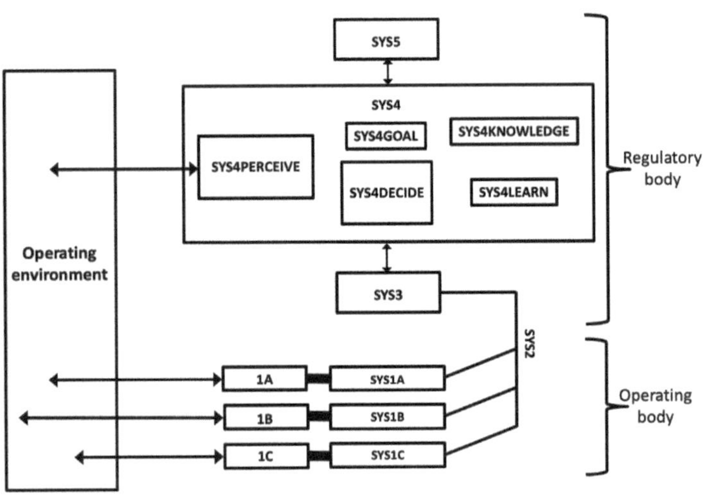

FIGURE 3.2.3 The Revised Viable System Model.

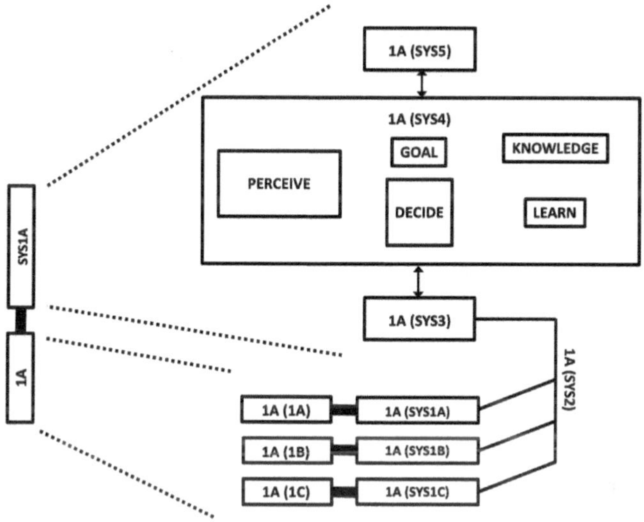

FIGURE 3.2.4 Recursive structure of SYS1 entities mapped onto general rVSM structure.

3.2.3.1 SYS1 – implementation

An organization may have many SYS1s. They are characterized as subsystems that often, but not always, interact with the environment outside of the organization and, in practical terms, fulfill the organization's function. In the context of a factory, there will be several SYS1 entities, for example:

- Reception and storage of raw material
- Manufacturing lines for different products
- Storage for manufactured products
- Sales teams
- Shipping

Depending on the organization's size, there may be many SYS1 entities worldwide. For a multinational organization, they may organize the SYS1 entities by country, i.e., all the above SYS1 entities are combined by location. Again, this brings us back to the concept of recursion; to be able to visualize the organization with sufficient resolution, we sometimes need to pack smaller entities away in the knowledge that we can always zoom in and look at those entities individually should the need arise.

Another aspect of recursion that applies to SYS1 in particular, is the idea that each of these SYS1 entities is in itself a viable system, i.e., they include the features of SYS2-5 as we are going to describe. A manufacturing line is not just a manufacturing line. It has people who know what is required of it, ensure that it is supplied with materials, oversee it as it operates, and can intervene directly if something goes wrong. Beer was keen to stress the idea that giving as much autonomy as possible to all SYS1 entities will lead to the most successful mode of operation. The idea is that for many things that can happen at the SYS1 level, the people on the frontline know better than someone in the boardroom how to address the problem. We will see later that autonomy does sometimes need to give way to other forces that might come into play, particularly when the system as a whole is under pressure, but, for the most part, SYS1 entities will operate most successfully if they are given as much autonomy as possible.

SYS1 entities (technically, subsystems) can be seen as essential, indivisible systems, much like the essential, indivisible systems that make up the subatomic baryons, some of the building blocks of matter. The functions assigned to SYS1 entities are usually those functions that actually do what the system as a whole is designed to do. It could be manufacturing, shipping, sales, or any one of many other functions essential to the operation. A manufacturer with no ability to sell its products will not be able to survive. Using Beer's original model for the VSM, he likened SYS1 to the vital organs of the body, each with their essential function. No matter how good the heart is at pumping blood, removing the lungs or the liver is still going to lead to death.

It is, of course, true to say that SYS1 entities can grow, shrink, or in some other way, change, but SYS1 entities are unique with regard to the rest of the organization in so far as failure of these subsystems is likely to be crippling for the organization. Elsewhere in the management structure, failures may occur and be dealt with in a way that doesn't have a significant impact on the rest of the organization, much like the plasticity of the developing brain allows for other brain areas to take over the functions of parts that might be damaged. However, the heart will never be able to take over the function of the liver, nor the lungs take over the functions of the kidneys. These are essential, indivisible systems in themselves.

3.2.3.2 SYS2 – communication

How does the SYS1 sales entity know what they are selling? How does a SYS1 manufacturing entity get its raw material? What happens if one manufacturing line in China is working at one speed and a line in France manufacturing a complementary part to that which is being made in China is working at a slower speed?

Although SYS1 entities are essential, indivisible systems that let the entire system fulfill its function by doing what needs to be done, there still needs to be coordination between the different entities. This is the role of SYS2.

While SYS2 will have some physical features (computers, telephones, paper records, etc.), it is essentially an informational system much like the spinal cord. For human beings, it is not the presence of a spinal cord that matters but rather the information transmitted through it.

Imagine SYS2 as an informational network that connects all the SYS1 entities. At any time, any SYS1 entity can access SYS2 to see what is happening in another SYS1 entity. More specifically, SYS2 is connected to each SYS1 at its local Regulatory Body (remember, each SYS1 is a viable system and has a Regulatory Body in the same way as the system as a whole has a Regulatory Body. For a SYS1 entity, this may be the supervisor of the production line or the head of the sales team).

The importance of communication rules in SYS2 cannot be overstated. SYS1 entities need to know that the purpose of SYS2 is coordination and not competition. For SYS2 to work effectively, the Regulatory Bodies of the SYS1 entities need to be confident that the unvarnished information that they are required to upload into SYS2 cannot be used in a punitive manner. If a SYS1 entity such as a manufacturing unit experiences technical problems with machinery for whatever reason, that information needs to be fed into SYS2 so that the other SYS1 entities can adjust their operation accordingly; the SYS1 entity that manages supplies and storage may request that a shipment of raw material is delayed because a storage backlog is anticipated and the SYS1 entity that manages sales may decide to contact clients and inform them about potential delivery delays.

If the SYS1 manufacturing entity decides to conceal the information that they are having technical trouble for fear that they may be penalized in some way when the problem eventually comes to light, disturbances have already rippled out across the rest of the Operating Body; the SYS1 entity that manages supplies and storage will have no more room to store the raw materials which have continued to arrive and the SYS1 entity that manages sales will have to deal with angry customers wondering where their orders are.

An Operating Body of SYS1 entities that uses SYS2 openly and honestly will create a SYS2 that communicates information and engenders trust.

SYS2 also has three other functions:

1. Transmission of information to SYS3.
2. Transmission of alert signals to SYS3, 4, and 5 (although this isn't the only route for alert signals). Alert signals are an important feature of the rVSM, and we will consider them separately later.
3. Provide reflex coordination of the Operating Body. Much like the neurological reflexes in humans that only occur in the spinal cord and do not require the input of the higher brain functions, SYS2 can provide reflex-based coordination of the SYS1 entities by accelerating or slowing certain SYS1 entities with respect to others.

In the language of systems, we can describe SYS2 as a requisite meta-system, i.e., it is essential for coordination. It requires energy to run, but without it, the SYS1 entities would not be able to coordinate their activities.

3.2.3.3 SYS3 – coordination

Suppose SYS2 is the information flow between SYS1 entities and the transmitter of this information into the Regulatory Body. In that case, there needs to be an entry (and exit) point in the Regulatory Body (comprising SYS3-5) that is connected to SYS2. That area is in SYS3.

SYS3 is a subsystem whose function is the coordination of activity between other subsystems. Its primary functions are:

1. Regulating the flow and the nature of the information within SYS2 to ensure that SYS1 entities can access the information that they need and can upload information that will be relevant to the rest of the SYS1 entities and the rest of the system as a whole
2. Stabilizing the relationship between the SYS1 entities where required (homeostatic and allostatic control)
3. Filtering SYS2 information and transmitting it to SYS4
4. Formatting and disseminating SYS4 output into SYS2 so that SYS1 entities can access it promptly
5. Receiving and distributing alert signals as they are received
6. Intermittent monitoring

SYS3 can operate in any one of three modes:

1. Monitoring mode – This should be the default mode for SYS3. When operating in this mode, SYS3 acts as both an information filter and an information junction box. It is sensitive to the information passing through it and monitors for any signals from SYS2 that suggest any of the SYS1 entities are not operating within normal parameters. This monitoring is passive as it is based on the information coming in from SYS2. Monitoring mode also has an intermittent submode that can run in addition to it:
   a. Active Monitoring submode – This active monitoring function is an important one. Originally referred to as an audit function in Beer's original model (he also called it System 3*), this function is independent of SYS2 and is an intermittently used two-way communication pathway between SYS3 and the individual SYS1 entities. Although we know that SYS1 works best when it is allowed to do so autonomously, the Regulatory Body also needs the ability to monitor intermittently what is actually happening in SYS1 entities. For example, it may be that an SYS1 entity is underresourced in time, energy, or materials and that this information has not been communicated through SYS2 for whatever reason. As such, the SYS1 entity has developed a workaround (an efficiency/thoroughness tradeoff), and now there is a difference between work-as-imagined (what the Regulatory Body thought was happening within this SYS1 entity) and work-as-done (what they are actually doing). This information can then be fed upward into SYS4, which

might then decide that the procedures should be adjusted to account for this more efficient way of operating.

2. Homeostatic mode – Beyond passive and, occasionally, active monitoring, SYS3 does have some autonomous measure of control over the Operating Body. This homeostatic control is an autonomic function of SYS3. Even without transmission of information upward to SYS4, SYS3 can detect where imbalances are happening within the SYS1 entities. If something in SYS1 is operating outside its normal parameters, SYS3 can detect this and decide on a course of action to restore balance. It may be that normal operation can be restored by reallocating some resources or temporarily slowing down the whole Operating Body. In this way, SYS3 can exert autonomic control over the Operating Body within parameters set for it by SYS4. To close the loop, SYS3 informs SYS4 that it has had to operate in homeostatic mode. The communication of this fact upward is important because if SYS3 is constantly operating in Homeostatic mode to try and maintain stable operation of the Operating Body, it is likely that there is something significant going on which will require specific attention from SYS4. It could be defined as the organizational experience of constantly "fire-fighting," i.e., something is out of its normal range here and must be fixed, then something else is out of its normal range over there and must be fixed. If this is the case, SYS4 may tell SYS3 to go into Allostatic mode and shift the parameters that define the normal operation.
3. Allostatic mode – As we described in Part 2, allostasis is the restoration of stability through change. A system that is constantly trying to use homeostatic control to keep itself stable is likely to be functioning poorly. If it is increasingly difficult to keep certain parameters within their expected range, it may be that the ranges need to be changed, if only temporarily. When SYS3 is in Allostatic mode, it adjusts the normal parameters expected from the SYS1 entities. These are transmitted via SYS2 so that all SYS1 entities can see what is going on and adjust their operation accordingly. Once these ranges have been adjusted and the Operating Body begins to function again, SYS3 will most likely return to Homeostatic mode to ensure that all SYS1 entities are now operating within the new parameters before reverting back to Monitoring mode. The effects of Allostatic mode may be quite significant, which is why it is regulated by SYS4. In Allostatic mode, one or more SYS1 entities may be reduced, shut down, or have their functions subsumed by another subsystem. For a large, multinational organization where the SYS1 entities are entire companies or regional branches of the larger company, an allostatic response could be shedding several of these SYS1 entities.

SYS3 can be viewed as the autonomic control center of the organization. If the principle of autonomy is working at the SYS1 level, SYS3 will have an easy time of it. All that it needs to do is coordinate the flow of information between the Operating Body and SYS4 and ensure that autonomous self-regulation is happening in the way that is expected. Again, by designing the system so that the SYS1 entities are as autonomous as possible and have full access to the information in SYS2, SYS3 may only very rarely need to go into Homeostatic mode and potentially never go into Allostatic mode.

3.2.3.4 SYS4 – decision

SYS4 is big. Even when Stafford Beer described System 4 of the original VSM, there was a lot to it. In the rVSM, there is going to be even more because this is where a lot of the new ideas that we covered in Part 2 are going to apply. As I was designing the layout for SYS4, I did think about breaking it up into separate parts, i.e., increasing the number of systems in the revised VSM from five to eight, nine, or ten. However, although SYS4 has a lot in it, all the separate components work in parallel to collect information, analyze it, and decide what to do with it. As such, I've kept SYS4 as the decision-maker of the rVSM but will describe the various functions that it embodies as clearly as I can using a specific nomenclature, which will be introduced shortly.

SYS4 in the rVSM is based on what we know about the architecture of the central nervous system. One of the most advanced architecture models was the one that we introduced in Part 2: ACT-R (Adaptive Control of Thought–Rational). ACT-R contains many functional modules that allow the human brain to carry out a wide range of activities and is shown diagrammatically in Part 2.

In the same way as Stafford Beer used the human nervous system as it was understood in the 1960s as the basis for the original VSM, I will be doing the same with ACT-R and adapting it based on some of the other concepts we covered in Part 2. I am hoping that this chimera of different areas of contemporary neuroscience and psychological research will describe more completely how SYS4 is meant to function.

Decision-making permits viability; the decision to fight or flee; the decision to react or not; the decision to feed, reproduce, travel, stay, build, communicate, sit, stand, etc. Every waking second of conscious human experience is defined by decision. Even doing nothing is a decision, even though it may also be thought of as our default position. However, before we get to the point of making a decision, we need to know what we are making a decision about.

In Part 2, we saw how the human stimulus/response process involves five steps:

1. Sensation – absorbing sensory information.
2. Attention – focusing only on that sensory information we see as currently relevant.
3. Perception – forming a mental model based on the sensory information we are focusing on.

4. Decision making – deciding how we should react.
5. Response – enacting our decision.

SYS4 includes steps 1–4. In Part 2, we also saw how humans have two different decision-making modes. It is worth going back to Part 2 and reviewing these two modes as understanding the difference between them and the relationship between them will be important. However, to summarize:

1. Mode 1 – this is the mode that we have inherited from our animal ancestors and allows us to make decisions in high workload situations. We can think of this as our instinctive mode of decision-making.
2. Mode 2 – the mode has evolved more recently and allows us to make more rational decisions using all the processing power of our highly evolved cerebral cortex. We can think of this as our rational mode of decision-making.

Mode 1 is always working. Information that comes into the brain is formed into a mental model, and based on that, Mode 1 decides if there is anything that needs to be decided immediately. Sometimes it will decide something immediately almost before "we" (our conscious mind) has realized that something is happening. Our hardwired neural programming is what allows Mode 1 to work. Mode 2 has to be consciously activated but, when it is working, it has the power to modify the output of Mode 1.

The five main functional subsystems of SYS4 are shown in Fig. 3.2.5:

Each of the subsystems contains further subsystems, and the modes of operation of SYS4PERCEIVE and SYS4DECIDE are dependent on each other.

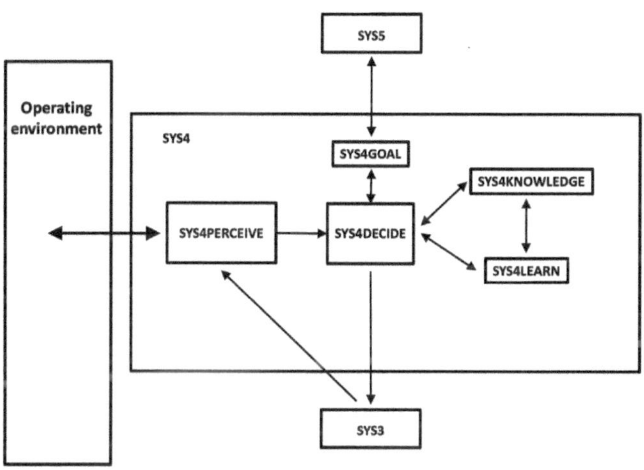

FIGURE 3.2.5 Layout of SYS4.

3.2.3.4.1 SYS4GOAL

This is the operational goal of the system as assigned by SYS5. Unlike the system identity embodied in SYS5, SYS4GOAL may change from time to time. For example, the operational goal may be to expand the operation to capture market share. Conversely, it may be to maintain operations how they are. Whatever the objective of SYS4GOAL is, that is what the rest of the system is going to be trying to achieve.

3.2.3.4.2 SYS4PERCEIVE

SYSS4PERCEIVE is always active. It is constantly building and rebuilding a model of the system and its operating environment and predicting how that model will change in the future. This is exactly the same phenomenon as human beings building the perceptual models that allow them to function in the real world. As we saw in Part 2, our reactions and decisions day to day are not based on reality but rather on our mental model of reality, which is built in the parietal lobes of our brain using sensory information. The perceptual model is usually sufficient to keep us oriented and functioning in our dynamic environment, but we do sometimes experience a mismatch. The act of reaching out to pick up a cup and accidently knocking it to the floor is an example of such a mismatch. An expanded view of SYS4PERCEIVE is shown in Fig. 3.2.6:

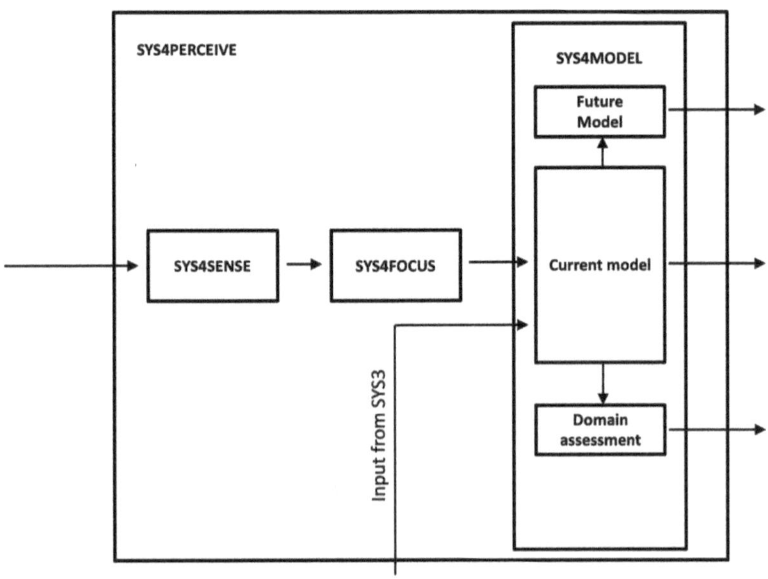

FIGURE 3.2.6 Layout of SYS4PERCEIVE.

The functions of SYS4PERCEIVE are:

1. To build a model of the state of the whole system and the state of the environment within which it is operating.
2. To project that model into the future.
3. To ascertain whether the current system and the expected state of the system in the future will satisfy SYS4GOAL, i.e., the operational activity is achieving the objective.
4. If it detects that the current or future system state may not satisfy SYS4-GOAL, it alerts SYS4DECIDE to respond to this mismatch.
5. Based on the current and future model, SYS4PERCEIVE can alert SYS4DECIDE and/or use the alert signal network to indicate a domain shift, e.g., a system that normally operates in the Complicated domain of the Cynefin framework is expected shortly to shift into the Complex domain. This will require a different mode of response.

To achieve these functions, much like the human nervous system, sensory information must be collected through SYS4SENSE. Organizations may use various techniques to collect information about market conditions, potential opportunities, changing global marketplaces, etc. To ensure that information relevant to the goal is emphasized, the information is filtered through SYS4-FOCUS, much like sensory information in humans is filtered through our attentional systems, including the reticular activating system. Like the human attentional system, SYS4FOCUS can operate in three modes:

1. Selective – actively focuses on a narrow part of the incoming sensory information.
2. Divided – focuses on two or more parts of the incoming sensory information.
3. Sustained – monitors the incoming sensory information to detect any signals that may be significant to SYS4MODEL.

Finally, SYS4MODEL uses this filtered sensory information to form a model. At this point, SYS4MODEL integrates the sensory information regarding the state of the Operating Body. This input comes from SYS3. It doesn't go through SYS4FOCUS as the model being built needs to reflect the state of the system itself to project the whole model into the future. Once SYS4MODEL has formed a model of the current state of the system and its operating environment, it tries to project this model into the future on the basis of not changing anything about how the system is operating. If this future model suggests that the system will still achieve the objective of SYS4GOAL, SYS4DECIDE doesn't need to do anything, i.e., "decide to do nothing." However, suppose sensory information suggests that things are changing in the operating environment or the system in an unanticipated way, i.e., some sort of disturbance. In that case, this mismatch between SYS4MODEL and SYS4-GOAL leads to the activation of SYS4DECIDE.

One final function of SYS4MODEL is to try and ascertain what operating domain the system is currently in and what future state it will be in according to the Cynefin framework. This is important because the mode of operation of SYS4DECIDE will be based on the Cynefin domain that the system is operating in. It becomes even more important if the disturbances have caused a domain shift. There is always the possibility that SYS4MODEL won't be able to make this determination. In the original Cynefin framework, this uncertainty about which domain the system is in is referred to as disorder, and the only option for a manager is to try and deconstruct the problem into smaller pieces or change their perspective on the system to see if working at a different scale might have a better chance of showing what domain the system is in.

3.2.3.4.3 SYS4DECIDE

In ideal conditions, SYS4DECIDE won't have to be active very often. Suppose SYS5 has assigned an appropriate objective to SYS4GOAL, and SYS4DECIDE has decided how best to achieve that goal and communicated that via SYS3 and SYS2 down to the SYS1 entities. In that case, there should be sufficient homeostatic and allostatic authority within SYS3 and sufficient autonomy in the SYS1 entities to achieve the system's function.

However, if SYS4PERCEIVE alerts SYS4DECIDE that the model of the current or future state of the system and its operating environment aren't going to satisfy SYS4GOAL, then something needs to be done. SYS4DECIDE is shown in Fig. 3.2.7 along with its associated subsystems:

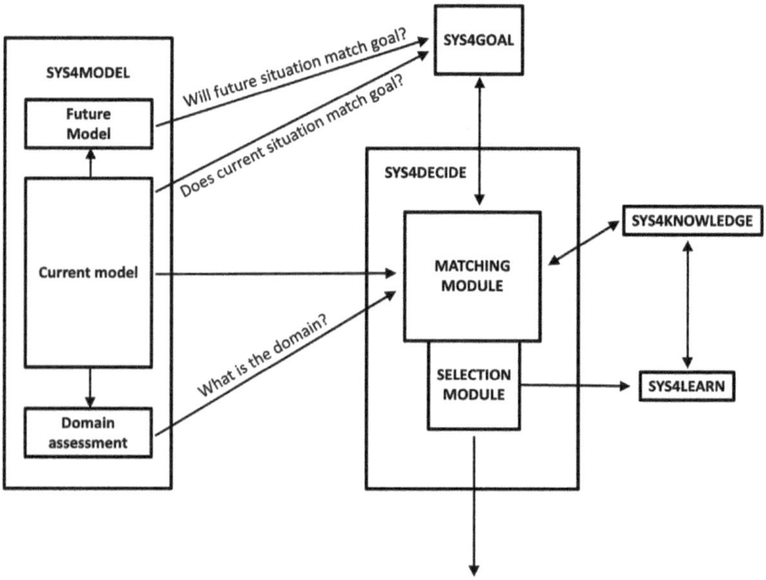

FIGURE 3.2.7 Layout of SYS4DECIDE.

SYS4DECIDE is modeled on the matching (trial) and selection functions of the ACT-R framework, which, in its turn, is modeled on how the basal ganglia of the human brain enable us to mentally trial possible responses to a particular stimulus before selecting one that we think will give the best result. We know already that human decision-making uses two modes:

Mode 1 – Instinctive. Always operating.
Mode 2 – Rational. Needs to be engaged consciously and can modify the output of Mode 1.

Wherever possible, SYS4DECIDE wants to be using Mode 2. However, there is a further complication to our nice bimodal model for SYS4DECIDE. Deciding responses for a whole organization is a much bigger deal than only having to decide something for ourselves. While organizations broadly have two options in how they make decisions, making a decision using the sort of processes that underpin Mode 1 should only happen in the most serious, rapidly changing, and chaotic circumstances, i.e., a domain shift into the Chaotic domain of the Cynefin framework.

This will make more sense when viewed in the context of the four modes of SYS4DECIDE. The mode that SYS4DECIDE will use is based on the Cynefin domain that the system is operating in or the Cynefin domain of the problem/subsystem that is the subject of the decision at hand. It may be that the system never drifts out of its normal domain, whatever that may be. In that case, an organization may find that it uses the same SYS4DECIDE mode most of the time. Broadly speaking, there are four modes that SYS4DECIDE can use to make decisions:

1. SYS4DECIDE/CATEGORIZE – This would be the decision mode when operating in the Clear domain (cause and effect relationships exist, and they are predictable) or when the issue to be decided is Clear. It is based on best practice and the sound knowledge that this particular problem requires this particular solution. The approach is sense-categorize-respond. If we were to equate this to one of Rasmussen's performance levels, this would be skill-based or rule-based behavior. This mode is shown in Fig. 3.2.8:
2. SYS4DECIDE/ANALYZE – This would be the decision mode when operating in the Complicated domain (cause and effect relationships exist but are not clear and so expertise is required) or when the issue to be decided is Complicated. It is based on good practice and the knowledge that this particular problem may have several solutions and that expertise may be required to select the best response. The approach is sense-analyze-respond. The level of performance being used is essentially rule-based. This mode is shown in Fig. 3.2.9:
3. SYS4DECIDE/PROBE – This would be the decision mode when operating in the Complex domain (cause and effect relationships are not clear and may not exist) or when the issue to be decided is Complex. It is based on iterative practice, i.e., probing the problem with experiments to see

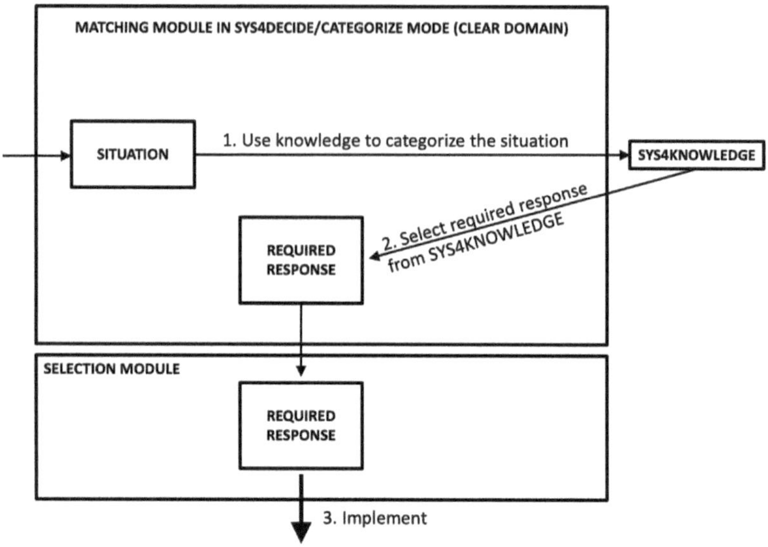

FIGURE 3.2.8 SYS4DECIDE/CATEGORIZE.

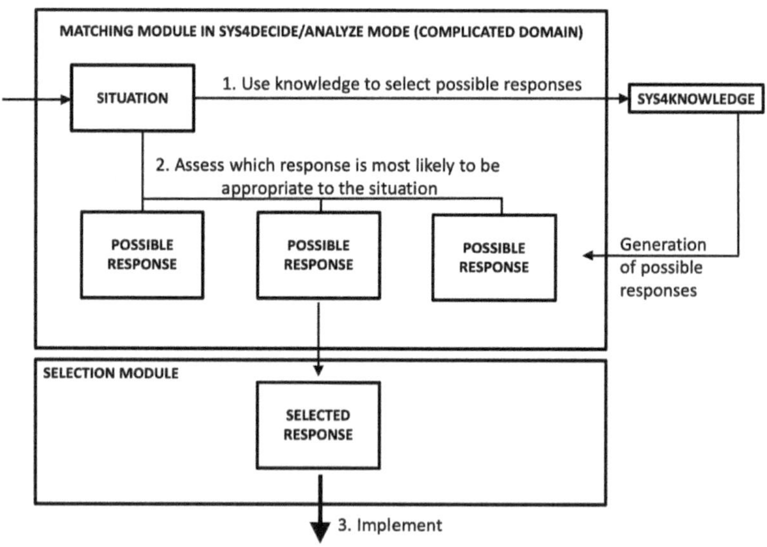

FIGURE 3.2.9 SYS4DECIDE/ANALYZE.

what responses work. If a response works in a small-scale experiment, it can be propagated to the rest of the system. If it doesn't work, we need to trial a different potential response. The approach is probe-sense-respond. The level of performance being used is Knowledge-based to create some Rule-based behaviors. This mode is shown in Fig. 3.2.10:

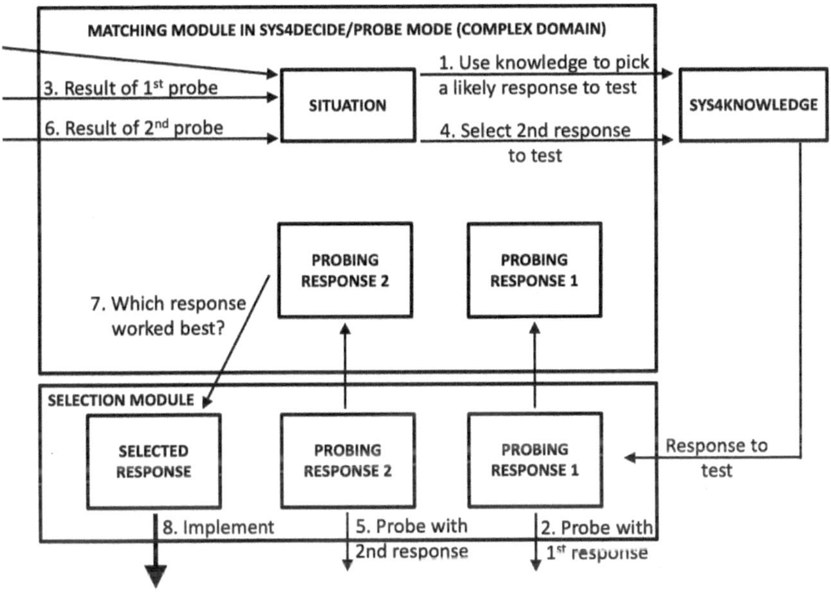

FIGURE 3.2.10 SYS4DECIDE/PROBE.

4. SYS4DECIDE/ACT-SENSE – This would be the decision mode when operating in the Chaotic domain or when the issue to be decided is Chaotic. A chaotic state is almost always the result of a domain shift. While a shift from the Complex domain to the Chaotic one can be highly disruptive, a shift from the Clear domain to the Chaotic one is disastrous. As we saw in Part 2, in the Cynefin framework, the boundary between the Clear and Chaotic domains is represented by a cliff, i.e., a shift from Clear to Chaotic is rapid and difficult to recover from. The first priority is to act to try and stabilize the system. There is no opportunity for careful analysis or experimentation. The approach is act-sense-respond. Like Mode 1 in human decision-making, we need to do what we think best to stabilize the situation. This Mode 1-style decision-making relies on heuristics, and for it to work, the organization needs to have some heuristics already stored up. These are courses of action that the organization can implement straight away that will potentially limit the effects of a serious domain shift. To use the examples of heuristics described in Part 2, the company needs to have access to courses of action that:

a. Have been used recently in a situation that is similar to this one (similarity matching).

b. Have been used frequently in other situations to limit damage (frequency gambling).

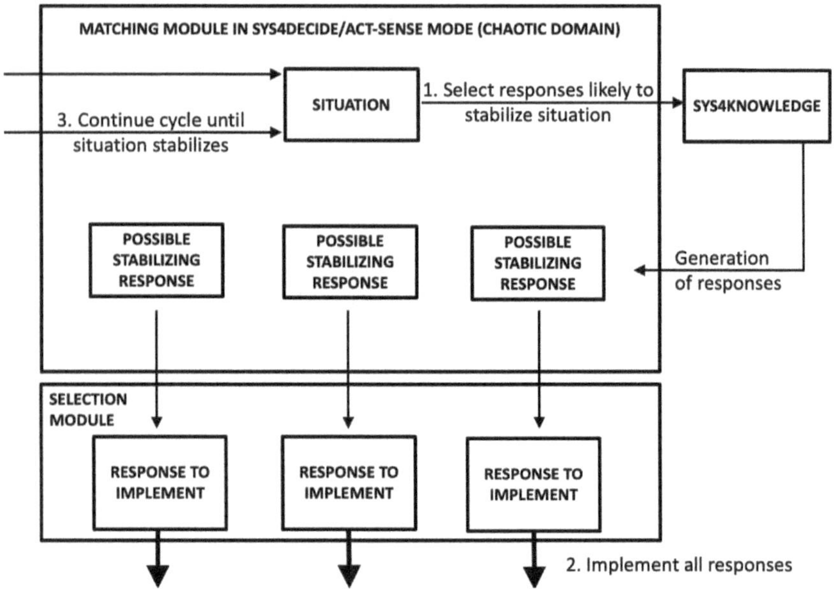

FIGURE 3.2.11 SYS4DECIDE/ACT-SENSE.

The level of performance being used is Knowledge-based given the novel and highly dynamic nature of the situation, although if heuristics are used to select rules that potentially may help limit the damage, this represents a mix of Knowledge-based and Rule-based performance. This mode is shown in Fig. 3.2.11:

Once SYS4DECIDE has settled on what it thinks is the most appropriate response, this decision is conveyed to SYS3 and then to the SYS1 entities for implementation.

3.2.3.4.3.1 SYS4DECIDE failure modes

Like human decision-making, SYS4DECIDE is also susceptible to different failure modes depending on what level of performance is being used to make decisions as well as any heuristics we are relying on:

1. Errors at the RB level – The production rule we have selected to deal with the current problem is inappropriate, i.e., the plan is wrong.
   - **a.** Similarity matching – We may overemphasize the similarity between two situations and so miss crucial differences that would mean that the production rule we're planning to use isn't going to work.
   - **b.** Frequency gambling – We have engaged in a statistical act of risk taking. We are gambling that a rule used frequently will have a higher chance of addressing the problem. This is not always the case.

   c. Overconfidence — Studies have repeatedly shown that we have more confidence in the accuracy of our knowledge and abilities than is justified. Overconfidence reinforces the similarity matching and frequency gambling programming that we have and increases the likelihood of making an RB error.
2. Errors at the KB level — The four main causes of KB errors are:
   a. Missing knowledge — There is no way to formulate an appropriate response because the information telling us how to do so is simply not there.
   b. Processing limitations — it may be that the situation is evolving so quickly that the individual can't maintain an accurate mental model on which to base their decisions.
   c. Oversimplification — In trying to eliminate complexity, we ignore some vital elements of the problem and so come up with an unsuitable plan.
   d. Overconfidence — in the same way as with RB errors, we generally have more confidence in the accuracy of our skills and knowledge than is justified, and this will reinforce any KB solutions we come up with even though they may be wrong.
3. Confirmation bias — We overemphasize situational features that confirm our mental model and tend to ignore important cues that would suggest that our model is wrong.
4. Plan continuation bias — We tend to want to continue with a plan once we have started to implement it even though the situation may have changed.

At an organizational level, these failure modes that affect human decision-making can be just as significant. Many of them are related to how knowledge is stored, and how easily accessible it is, particularly in time critical situations. To try and mitigate the risks that inappropriate responses are selected, the connection between SYS4DECIDE and SYS4KNOWLEDGE needs to be robust, and the way in which information is organized in SYS4KNOWLEDGE needs to ensure that the required information is readily accessible.

3.2.3.4.4 SYS4KNOWLEDGE

Our basal ganglia retrieve production rules from our procedural memory to mentally try them out before selecting the one that it calculates has the highest probability of achieving the goal. Similarly, SYS4DECIDE accesses procedural knowledge stored within SYS4KNOWLEDGE to find potential responses. These are then tested within SYS4DECIDE to try and assess which response is likely to work best. SYS4KNOWLEDGE does not just represent the knowledge encoded in operating manuals or software. It also comprises the procedural knowledge of team members and experts. SYS4KNOWLEDGE is not just the physical system of stored knowledge but rather the informational system of knowledge in its many physical and biological forms across and sometimes outside of the organization.

Much like the subdivisions in human knowledge storage, SYS4KNOWLEDGE also contains declarative knowledge in addition to procedural knowledge. Declarative knowledge is raw information rather than production rules that describe how things are done.

As we saw with the phenomenon of entropy, compartmentalized knowledge must be maintained, and this maintenance requires energy. It may be that to keep organizational responses agile, cycles of practice may need to be introduced so that the procedural knowledge that might need to be deployed quickly in the future is readily available and quickly accessible in SYS4-KNOWLEDGE. This is the principle behind training. It also means that SYS4KNOWLEDGE is depleted when employees leave an organization. Although it cannot be easily quantified, the knowledge stored in employees' brains may be vital to the system's operation as a whole, particularly in systems where there is a large gap between work-as-imagined and work-as-done. Additionally, maintaining SYS4KNOWLEDGE can help mitigate the risks associated with the SYS4DECIDE failure modes.

3.2.3.4.5 SYS4LEARN

The success or failure of a response selected by SYS4DECIDE is a useful data point. For an organization to learn, there needs to be a way of including data about what actually happened when a response was selected, communicated to SYS3, and implemented at the SYS1 level. For example, suppose the system is operating in the Complicated domain, and so SYS4DECIDE/ANALYZE was the SYS4DECIDE mode. In that case, if the selected response was successful and SYS4GOAL is now being satisfied, this will mean that the next time this mismatch occurs, we will have to do less analysis before we select this response again. If the response continues to work reliably, it may be that future decisions are made in the Clear domain, i.e., sense-categorize-respond. This is learning process is akin to James Reason's assertion that expertise means never having to resort to KB processing because enough rules have been learned to allow for most problem solving to happen at the RB level.

SYS4LEARN is also active in trying to expand SYS4KNOWLEGDE. When things have gone well, even though SYS4 may have had no part in the success (e.g., it was dealt with autonomously at the SYS1 level), SYS4LEARN tries to understand how this success was achieved in the face of dynamic conditions so that it can save that knowledge and promulgate it in order that it may be of use in other parts of the system. In this way, SYS4LEARN is linked to SYS3, especially when it operates in its active monitoring submode (i.e., similar to the audit function of the original VSM). It is in this mode that SYS3 and hence SYS4LEARN can identify any gap between work-as-imagined and work-as-done.

The connection between SYS4LEARN and SYS4KNOWLEDGE is crucial, and it is an area of system design that has the potential to significantly increase the capabilities of the overall system as we will see later in some of the case studies.

3.2.3.5 SYS5 – identity

SYS5 is a relatively small but essential part of the overall system. In the original VSM, Beer described System 5 as being the boardroom level expression of the organization's goal and what the company ethos should be while trying to achieve it. SYS4 would make day-to-day decisions in the organization unless circumstances determined that SYS4 didn't have the scope to achieve the goal determined by SYS5 while operating within the ethos set by SYS5. It would only be at this point that SYS5 would need to become involved to decide if the goal should change or the ethos should change.

Rather than using the word "goal" or the word "ethos," I have described SYS5 as identity. Anything that is not readily interchangeable with something else that is similar has an identity. A neuron in your central nervous system is unique in how it works internally and in the way it is connected to other neurons. We cannot readily swap one neuron for another and expect everything to function in exactly the same way as before. Individuals are the same in terms of how they fit into a team or a department.

Systems have identities defined by their function and their ethos. The function is what the system is designed to do, and the ethos is, in broad terms, how it intends to do it. From a systems perspective, ethos is a harder word to define. It determines what can and can't be done to fulfill the system's function. It may be the case that changing conditions mean that for an organization to survive, the ethos needs to change. This, again, is an allostatic response.

However, ethos has another meaning that is difficult to explain solely by reference to systems engineering. Ethos is the character of the organization in much the same way as we would talk about the character of an individual. It describes the values that SYS5 says are important to the organization, not just in terms of how successfully it accomplishes its function but also by what means it accomplishes its function. Ideally, this template of behavior should flow down through SYS4, SYS3, and out to all the SYS1 entities.

Some systems experts might say that ethos is an emergent property of the system. Aside from my doubts about the idea of emergence as a real phenomenon in its own right, I don't agree that it somehow materializes out of nothing.

If a new organization is designed from the ground upwards using the rVSM, SYS5 would, by necessity, have to describe the function of the organization. At the point of inception, the ethos may be to fulfill the function "by any means necessary," i.e., do whatever it takes to carve out a market share. As the organization grows, it will be shaped by the people within it, and so as people rise through the ranks, the ethos may change.

Conversely, an organization that is failing may bring in new board members, a new chairperson, and a new Chief Executive Officer. They may well conclude that the organization is failing because the ethos that SYS5 is projecting is not suitable. The case study we will look at later concerning Continental Airlines is an excellent example of this.

An organization may not know that it has an ethos, but one will develop simply by existing and being active. It is at once mutable, directable and, potentially universal in how it affects the rest of the system, which is why it is an important part of SYS5 and is worthy of careful attention.

3.2.3.6 Alert signaling network

The alert signaling network is an independent system that is connected to every SYS1 entity and to every system in the Regulatory Body. So as not to overcomplicate the diagram of the rVSM, the alert signaling network is not shown as it would involve linking up every subsystem with every other subsystem.

The alert signaling network is similar to SYS2 but is independent of it, although SYS2 should show the alert signal if the system is activated. In the original VSM, Beer referred to this as algedonic signaling, literally pleasure/pain signaling. The principle of the alert signal is that any component in the system can get the attention of every other component if they think an issue is important enough. It may be that they see an oncoming, serious problem that hasn't been planned for, or it could be because they identify a really good opportunity to take advantage of. The idea of the alert signal is that it is used rarely, but it mobilizes resources quickly when it is used.

If the alert signal were being used all the time, it would suggest that something is seriously amiss, probably in SYS4, and a degree of allostasis is required to reconfigure the system to account for this persisting issue and for why it wasn't communicated via SYS3.

The alert signaling network can work in any direction and from any source. It can alert the whole organization or alert one component by bypassing the normal SYS2 channel. Under normal circumstances, a SYS3 that functions well should be able to detect trouble in the Operating Body by having a sensitive filter to what is coming into it from SYS2. If this is the case, the alert signaling network may never be used.

I've referred to this element of the rVSM as a network rather than a system. With SYS1-5, the connections are specific and sequential, e.g., SYS2 feeds into SYS3, which feeds into SYS4, etc. The idea of a network is that it links everything in the system in a nonsequential way so that anyone can contact anyone immediately if they think the information they need to convey is important enough. As well as having a physical arrangement, people also need to feel confident and empowered to use it when they feel it is necessary. This could have made a difference in the two Space Shuttle disasters that we will look at in the next part of the book.

3.2.3.7 Transducers, amplifiers, and reducers

These concepts were part of the original VSM, and they are just as important in the rVSM. They are subtle functions that different parts of the system might possess, but they will impact how successfully the system manages information and reacts to increasing or decreasing complexity inside or outside of the system.

If we consider SYS4SENSE, the amount of data that will be available to it from the outside environment may well be vast. For SYS4SENSE to work, much like our physical senses, it needs to turn this raw sensory information into something that the rest of SYS4 can use. All our senses turn external stimuli into electrochemical impulses, which can then be processed further by the nervous system. In the rVSM, a transducer is anything that turns one type of information into another type. The reason that transducers are important is that the fidelity of this conversion can affect the results of processing downstream. Knowing that a system relies on transducers means managers need to check that the transduction process is accurate. Transducers shouldn't be editing or simplifying information because, in the case of SYS4, that would be a function of SYS4FOCUS and SYS4MODEL.

In the original VSM, Stafford Beer also talked about variety amplifiers (a "variety amplifier" is an element of the VSM) and reducers. This was a way of saying that systems should have a way of increasing their complexity or somehow decreasing the complexity of their operating environment. To go back to Ashby's law, we know that the complexity of a system needs to be greater than the complexity of its operating environment:

> When the variety or complexity of the environment exceeds the capacity of a system (natural or artificial), the environment will dominate and ultimately destroy that system.

In the original formulation, variety is a term that denotes how many states a system or the environment can take. For a system to be able to manage the variety possible in the environment, it must be able to exhibit at least the same degree of variety or even more.

We also said that Ashby's law could be stated a bit more simply:

> The more we expect our systems to do and the more resilient we expect them to be in the face of disturbances exerted by the dynamic environment in which they function, the more complex the system must become.

How then can a system react if its operating environment suddenly becomes more complex? Does it need to be completely redesigned to avoid the fate of being "dominated and ultimately destroyed" by its environment? Beer would say, not necessarily.

In managing changes in complexity either internally or externally, we can employ complexity amplifiers and complexity reducers. If environmental complexity has increased, we can increase complexity in our system to match that. For example, recruiting more humans into the system can harness their innate flexibility to deal with this increase in complexity. Increasing SYS4-KNOWLEDGE introduces more information into the system, which may be sufficient to handle the changes happening in the external environment. If the changes are temporary, it may be enough to outsource certain functions to other organizations that are more readily able to engage their resources to deal with the problem. These solutions can then be filtered back into the system.

Alternatively, rather than increase the system's complexity, we can use complexity reducers to decrease the effective complexity of the environment. For example, instead of trying to operate the system based on ensuring that the future projections of SYS4MODEL are likely to have their goals met, management may decide to shorten the time frame which the system is working to. It may be necessary to reduce the complexity required to plan into the future to better handle the dynamic disturbances happening today. It may also be possible to reduce the high complexity of the environment by aggregating and averaging conditions. For example, rather than trying to account for the needs of individual clients, should the number of those clients or the number of possible needs increase, the system instead tries to account for the *average* requirements of groups of clients. This is the basis of grouping consumers into classes based on preferences, socioeconomic background, etc. It is easier to manage these classes of consumers rather than all the consumers in all the classes individually.

Complexity amplifiers and reducers need to be used with caution. While they have their benefits, they also have their drawbacks. In some ways, their use represents an allostatic way of dealing with disturbances; they involve dynamical changes that will persist until they are reversed. Suppose complexity amplifiers and reducers are being used constantly. In that case, it may be a sign that something more fundamental needs to be changed in the system to more permanently account for the changing environmental complexity.

Chapter 3.3

Capability and adaptive capacity

It is not the strongest species that survive, nor the most intelligent, but the ones most responsive to change.

Charles Darwin, naturalist

Chapter outline

3.3.1 Capability

Structure is designed. Function is assigned. From the highest level of the rVSM, the function of a system, subsystem, or component is what we need it to do. We know that an organization will need certain departments that are related to its overall purpose, and so those departments are assigned specific functions. The process of assigning functions will run interdependently with the process of designing the structure of the system, as described in the previous chapter.

Once structure and functions are mapped out, we next need to consider the capability of the components. Do the components have access to the necessary resources and sufficient interconnectedness with the rest of the system to carry out their functions? Erik Hollnagel developed a modern analytical approach to complex sociotechnical systems where all the functions within the system were considered in terms of their interrelatedness.[1] To fit in with the nomenclature being used in this chapter, I have adjusted the language that Hollnagel uses to describe aspects of his Functional Resonance Analysis Method (FRAM). For example, the basic elements in the FRAM are functions. Using our nomenclature, the basic elements in our complex system are the components that have an assigned function and may or may not have the capability to carry out this function.

The FRAM tells us that six parameters determine the capability of a component to fulfill its function:

Designing and Managing Complex Systems. https://doi.org/10.1016/B978-0-323-91609-7.00010-X

- Input – The availability of the matter, energy, or information that the component is meant to use. This includes the channel through which the matter, energy, or information arrives.
- Output – The onward path of the changed matter, energy, or information that the component has used. This includes the channel through which the matter, energy, or information leaves.
- Control – How the component is monitored and controlled locally and how that local control is coordinated with respect to the rest of the system.
- Preconditions – The conditions that must exist for the component to be capable of fulfilling its function.
- Resources – The resources that must be available.
- Time – The time that must be available. This element is often overlooked as organizations attempt to increase productivity.

The capability of subsystems and components to fulfill their assigned functions is determined by how adequately these six parameters are addressed. For example, increasing the output expectations for a component/subsystem without adjusting the time or resources available may lead to component failure.

3.3.2 Anticipating system dynamics

So far, in building our system, we have designed a structure of subsystems and components, and we've assigned functions to them. We've also ensured that they all have the capability to fulfill those functions. Does this mean that we are ready to commence operations?

We need to take one more step before firing up our shiny new system: we need to consider how it will *actually* run when it is operational.

So far, we have built a system based on our idea of what it needs so that it can do what we want. In a way, we have identified the minimum gene set for the type of organization we want to build, and we've put those building blocks in place. The question is, will the system be viable? If it is viable, how long will it be viable for, and what happens when it experiences its first unexpected disturbance?

Beyond the parameters that need to be addressed to allow a component/subsystem to fulfill its function, there is another characteristic of the component/subsystem that needs consideration. Chances are we have assigned functions and ensured capability for a system that is operating under normal conditions. How is it going to operate in nonnormal conditions?

The transition from a normal to a nonnormal condition may be gradual, or it may be rapid. A nonnormal condition could be caused by external disturbances such as an international financial crisis, a pandemic, or a new market competitor. A nonnormal condition could also be caused by an internal disturbance such as an IT breakdown, a staffing crisis, or a significant error occurring in some part of the organization.

How the system reacts to disturbances will determine viability as much as the structure and functions we have designed. Depending on the nature of the disturbance, it may be that the components and subsystems that make up the system as a whole need to start working in a different way than they were originally designed to. This is more than a local homeostatic response in a SYS1 entity. This is homeostasis (and possibly allostasis) on an organizational scale. When we considered homeostasis and allostasis at the scale of subsystems and components, this type of autonomic control was understandable because some sort of regulator was managing these changes (SYS3 and possibly SYS4). For homeostatic/allostatic change on a much wide scale, how will it be regulated?

This is a fundamental problem with systems designed in a semihierarchical manner: when a local function isn't being carried out adequately, the next level up needs to intervene and help the component/subsystem return to within normal operating parameters. An external or internal disturbance that is significant enough can spread through an organization like ripples on a pond. Every aspect of the operation can be unbalanced, and the normal hierarchical concept of monitoring subsystems and intervening when necessary won't be sufficient.

There needs to be another characteristic built into the system that can be deployed at the local level but also extend outward from the local area, linking up across the system so that all areas can react to the effects of the disturbance that are propagating throughout the system. This characteristic is adaptive capacity.

3.3.3 Adaptive capacity

Adaptive capacity is a characteristic that is difficult to describe in the context of the broad areas of system design (structure, function, and capability) we have already covered. In some respects, you might say that it is a capability, but it works differently from capability as we have defined it. We've said that if the six parameters required for a component/subsystem to be able to carry out its assigned function have been adequately addressed, then that component/subsystem can fulfill its function. Capability, in this sense, is a local phenomenon.

The adaptive capacity of a component/subsystem is like the capability to adapt due to local disturbances AND its ability to coordinate with nonlocal components/subsystems to coordinate this adaptation. In a similar way as the alert signaling network is distinct from SYS2 because of its nonhierarchical arrangement, adaptive capacity embodies both local change and system-wide change in a networked manner.

David Woods, a renowned professor of cognitive ergonomics, described the phenomenon of adaptive capacity more extensively in a series of papers. He defines it as the potential for modifying what worked in the past to meet challenges in the future; it is a relationship between changing demands and responsiveness to those demands relative to goals.[2]

What follows is a summary of some of these ideas though slightly adapted to fit in with the nomenclature that has been used in this and the preceding chapters.

Woods states that there are two universal truths that mean that adaptive capacity is important for systems, particularly those in the Complex domain:

1. Resources are always finite.
2. Change is constant. This idea of change reflects both natural changes in how a system evolves and the external/internal disturbances it might encounter.

Woods presents 10 observations about the adaptive capacity of components/subsystems that relate to their functioning within a system.[3] By recognizing and accounting for these observations, he believes that we can endow a system with enough adaptive capacity to exhibit "graceful extensibility" in the face of changing demands. This is a way of describing how a system gracefully and in a coordinated way expands to absorb disturbance to stop it from pushing the operating point of the system toward a failure boundary. These observations fall under three different themes:

A. Managing the risk of saturation
B. Networking the adaptive capacity of components
C. Outmaneuvering constraints

We will look at these three classes of observations and give some examples of how these behaviors may manifest themselves in the real world. They may appear a little confusing at first, but once you have read them a few times, you will start to see how they relate to each other.

A. Managing the risk of saturation
   1. The adaptive capacity of any component/subsystem is finite. Therefore, all components/subsystems have bounds on their adaptive capacity.
   2. Disturbances can and will occur that will challenge the adaptive capacity of any component/subsystem. These disturbances will demand a response; otherwise, the component/subsystem is subject to failure.
   3. All components/subsystems risk saturation of their adaptive capacity. Therefore, components/subsystems require some means to modify or extend their adaptive capacity to manage the risk of saturation.
B. Networking the adaptive capacity of components
   4. No single component/subsystem, regardless of level or scope, can have sufficient adaptive capacity to manage the risk of saturation alone. Therefore, alignment and coordination are needed across multiple interdependent components/subsystems in a system.

5. Neighboring components/subsystems in the system can monitor and influence (constrict or extend) the adaptive capacity of other components/subsystems to manage their risk of saturation. Therefore, the effective range of the adaptive capacity of any set of components/subsystems depends on how neighboring react as the risk of saturation increases.
6. As components/subsystems pursue their goals, they modify the pressures experienced by other components/subsystems. This leads to trade-offs as the different components/subsystems constantly try and find their best operating point.

C. Outmaneuvering constraints

7. The performance of any component/subsystem as it approaches saturation differs from that of that component/subsystem when it operates far from saturation. Therefore, there are two fundamental forms of adaptive capacity for components/subsystems to be viable: base and extended. Both are necessary but interconstrained.
8. All adaptive components/subsystems are local (constrained based on their position relative to the world and relative to other components/subsystems in the system). Therefore, there is no best or omniscient location in the system.
9. There are bounds on the perspective of any component/subsystem; the view from any point of observation at any point in time simultaneously reveals and obscures properties of the environment and the system. This limit is overcome by shifting and contrasting over multiple perspectives.
10. There are limits on how well a component's/subsystem's model of its own and others adaptive capacity can match the reality of the extent of their adaptive capacity. Therefore, miscalibration is the norm, and ongoing efforts are required to improve the match and reduce miscalibration.

We can tie these ideas back into the rVSM by saying that increasing the adaptive capacity of components and subsystems increases the overall variety of the system. According to Ashby's Law of Requisite Variety, the more states that the component/subsystem/system can have, the better it can manage complex, dynamic situations.

Not every type of component can have an innate adaptive capacity. A machine, for example, is relatively limited in its adaptive capacity. It can do what it has been designed and built to do. The source of adaptive capacity within complex sociotechnical systems is usually the human components. For clarity, and because adaptive capacity in this type of system component is most significant, I will continue exploring adaptive capacity only with reference to human performance.

Woods' 10 observations about adaptive capacity are a starting point for us as system designers and managers. We need to ask ourselves a final question before starting up our newly built system. Does my system have enough adaptive capacity to deal with the unexpected (and, sometimes, expected) disturbances, both external and internal, that it will experience when it starts real-world operations? We know that components/subsystems have base adaptive capacity. Knowing what we know now about their behavior under pressure, what can we do to ensure that they will demonstrate sufficient extended adaptive capacity when the situation demands?

The unfortunate answer is that it is difficult to give a perfectly matched solution to every potential problem that could be caused by insufficient adaptive capacity. So varied and numerous are the potential failure modes and so diverse the components/subsystems that it would be meaningless to try and present a list of all the possible ways of endowing systems with tools to help them deploy extended adaptive capacity when faced with disturbances. However, from the observations of Woods and from our knowledge of the normal dynamics of systems and how other systems in nature deal with disturbances, we can propose some principles based around Woods' three classes of observations:

A. Managing the risk of saturation:
Saturation alerting – As we will see later, one of the failure modes of complex systems is decompensation. In biological systems under pressure, the organism may appear normal when, in fact, it is engaging all its compensatory mechanisms to sustain normal function. However, there will come the point when the compensatory mechanisms are either exhausted, or the pressure is so great that the organism suddenly decompensates and loses viability. This happens in humans who have liver failure and those who are losing lung function. People with poor lung function can look remarkably well until you test their blood, and the results tell you that they have very low oxygen levels. This is relevant because to manage the risk of saturation, we need to alert others that we are becoming saturated. If we keep trying to compensate without indicating that we are doing so, we will reach the point where we rapidly decompensate and could take the rest of the system down with us. Saturation alerting is a way of alerting neighboring components/subsystems that you may need their help in extending your adaptive capacity and that they may have to extend their adaptive capacity to assist.

B. Networking the adaptive capacity of components
Coordinating responses – Another failure mode in complex systems is working at cross purposes. When a disturbance is detected, one set of components/subsystems may try and extend their adaptive capacity in one way to absorb the disturbance, and another set may try and extend in the other direction. Coordination is required to ensure that the whole system's adaptive capacity is extended in the right direction and in a synchronized

way to avoid the risk of amplifying the disturbance by working at cross purposes.

C. Outmaneuvering constraints
Partially decentralizing control – Observations 8–10 reflect the difficulty of extending adaptive capacity when you don't have the full picture of what is going on. If SYS1 entities were given complete autonomy in every sense, the response to a disturbance would be an uncoordinated mess. Alternatively, if the regulatory center tried to maintain the normal status quo of transmitting information from the SYS1 level, via SYS2 and then to SYS3 for filtering and up to SYS4 for decision, by the time SYS4 selects a response, the situation at the SYS1 level may have changed. Under normal circumstances, constraints are good because they set an expected limit on functional performance, and if those limits are exceeded, resources can be recruited to bring performance back into the acceptable range. However, if constraints are being exceeded in many different areas of the system because of a disturbance, expanding or, in some cases, eliminating the constraints may increase the ability to extend adaptive capacity at the SYS1 level. In this scenario, more decisions are being taken at the SYS1 level, and SYS2 and SYS3 can act in a monitoring and communicating capacity to try and synchronize the activities of the SYS1 entities. This is an important point; it is not directing activities but is instead acting as a conduit for information. Depending on the extent of the disturbance, the only response that could be enacted in time may well come from the SYS1 level. In this way, SYS1 is deciding things autonomously in an act-sense-respond manner similar to SYS4DECIDE if it is having to operate in the Chaotic domain of the Cynefin framework. The challenge then is to try and coordinate this partially decentralized control as well as possible. We will go into this type of behavior when we look later at a case study about Wal-Mart and its reaction to Hurricane Katrina.

Now that we have ensured that all of our components/subsystems have the capability to fulfill their functions and have given them sufficient adaptive capacity, we can start to look at how we can coordinate structure, function, capability, and adaptive capacity to give our system resilience.

Chapter 3.4

Engineering resilience

Science can amuse and fascinate us all, but it is engineering that changes the world.

Isaac Asimov, writer

The ideal engineer is a composite. He is not a scientist, he is not a mathematician, he is not a sociologist or a writer; but he may use the knowledge and techniques of any or all of these disciplines in solving engineering problems.

Nathen W. Dougherty, civil engineer

Chapter outline

3.4.1 Introduction

In the last two chapters, we've built on the idea that sociotechnical systems have certain features that increase the likelihood of being viable. Using successful examples of self-organization and viability from different fields, we've seen how structural, functional, and capability characteristics can be built into systems at the design stage.

Capping all of this off was the idea of adaptive capacity. We have seen that components and subsystems have a base adaptive capacity that allows them to fulfill their function under normal circumstances, adjusting to the slight changes in requirements that will naturally occur in any working system. We've also seen that to respond to larger disturbances, whether they come

Designing and Managing Complex Systems. https://doi.org/10.1016/B978-0-323-91609-7.00011-1

from inside or outside of the system, we need to consider the extended adaptive capacity of components and subsystems. The previous chapter looked at the features that allow extended adaptive capacity at the subsystem level to synchronize to give the system the ability to absorb the disturbance to prevent it from affecting the overall function. If we can endow our systems with this degree of adaptive capacity, we can start to consider them to be resilient.

As we will see shortly, there is more to resilience than adaptive capacity, though when we start considering these phenomena when viewed at the systemic scale, there is a necessary degree of overlap; if we expect a systemic response, we need to expect some coordination between the entities within the system that are generating this response.

As we delve further into this idea of resilience, we will be moving away from ideas about structural and functional elements of systems and will start focusing more on the dynamics that these systems exhibit and how we can get these dynamics to work in our favor. Some of the principles of Resilience Engineering do link directly to structural, functional, or adaptive characteristics of viable systems that we have covered in the previous chapter. However, some of the principles will be new and may only make sense in the context of whole-system responses to disturbances.

3.4.2 A note about terminology

It may be that some of you reading this are familiar with the idea of resilience, Resilience Engineering (RE), and/or operational resilience already. As someone who has been involved in this field from soon after it became recognized as a discipline in its own right, I can sympathize with those of you who sometimes feel overwhelmed by the terminology and, in some cases, the lack of agreement even about the definition of terms and phrases. Unlike the word "consilience," which is not in common parlance and so can easily be defined in the terms that Edward O. Wilson wanted, the word resilience is commonly used in everyday language, and so it becomes even harder to demarcate Resilience (with a capital letter) as something different. It is harder still that we can't see capital letters when speaking or listening! It is clear from my experience that this loose definition is having an impact on how the topic is being understood in operational environments and that there is certainly some crossover happening between the dictionary definition and the academic definition so that people being introduced to the subject aren't entirely sure where the boundary lies.

However, we must accept that it takes a while for the vocabulary and syntax to settle into a (reasonably) concrete form in any developing field. Read any paper about resilience, and you will see the term defined in a slightly different way. For clarity, I am going to use the following definitions:

- Resilience – the ability of a system to adjust before, during, and after expected or unexpected disturbances to maintain its function and take advantage of opportunities

- Resilience Engineering – the design and management processes that enhance a system's ability to be Resilient

3.4.3 Resilience Engineering

RE began to emerge as a subject shortly after the turn of the millennium.[1] Based in no small part on Charles Perrow's observation that the complex sociotechnical systems within which we work were becoming so complex that we couldn't reasonably predict where success and failure come from, a group of researchers in industrial engineering decided to see if they could find ways of engineering resilience into the systems that were being built and retrofitting it to systems that already existed.

It turned out to be a fruitful area of research. However, perhaps because of the ubiquity of the word resilience in common parlance, in some environments, the term became a catch-all expression for anything good about a system, i.e., "this makes us resilient." The ideas encapsulated in RE have also been seen under the umbrella term of Operational Resilience. Both are valid, but I think the word Engineering is valuable here if only for the reasons alluded to in the quotes at the start of this chapter: engineering is an active process, and we need to do it only on the basis of solid, logical principles.

One of the most appealing aspects of RE is that it focuses on things that go right. Normally in industrial engineering, if something goes wrong, however rarely, we spend a great deal of time and money attempting to understand why it went wrong. An RE practitioner would say that, yes, we need to understand what went wrong, but we also need to understand all the occasions when a negative event did not occur despite the presence of disturbances.[2] Those working in the system at the time were not necessarily working in ideal conditions but somehow managed to deal with the disturbances to preserve system function and have a positive outcome. With this in mind, RE suggests that there is much that we can learn from positive practice rather than focusing all our analytical efforts on occasions when things have gone badly. This is not to say that these occasions do not require further examination. It is just to say that they don't give us the whole picture of how systems behave under pressure.

Given that the humans are the components that exhibit the greatest variability in their performance in a sociotechnical system, it is often the activity of these humans that can potentially generate disturbances that occur within the system. As we've seen in the previous chapters, humans must deal with a wide variety of conflicting goals and often must make trade-offs to fulfill their functions in the time available. Is it any wonder then that, occasionally, these trade-offs do not pay off? Conversely, this same variability in those same humans gives them the adaptive capacity that allows them to potentially act as the buffer that can absorb a disturbance and allow the system to continue to function. The archaic view of this was that when things went wrong, the humans were always to blame. The RE perspective is that while the variability of human performance may be a source of disturbance in a system, the adaptive capacity of humans is the system's strength. Thus, we have one of the fundamental challenges in RE: how to manage variability in performance while still leaving enough flexibility to allow humans to deploy their adaptive capacity when needed. It is a dilemma with no easy solutions as it will always be a balancing act between constraining behavior to manage performance and loosening/removing those constraints on behavior to allow for flexible, adaptive responses when required.

It is true to say that RE grew out of the domain of Industrial Safety and was initially seen as a way of improving operational safety. The early research and conferences generally involved people from safety-critical industries such as aviation, healthcare, and nuclear engineering. As time went by, it became clear that the techniques that were being discussed to improve operational safety could also be used in industries that were not safety critical to improve agility and responsiveness. The conferences soon had more delegates from the worlds of software, web design, and commerce, and the principles of RE started being discussed in terms of commercial viability and business development. This may be another example of consilience at work: the analysis of system behavior and ideas for improving it that originated in one particular field were able to be translated successfully into other fields.

3.4.4 Resilience and the operating point

I've always found that a useful way of looking at RE is to do so in the context of Rasmussen's Dynamic Safety Model as described in Part 2. Although this was originally designed as a way of understanding how safety varies in sociotechnical systems, we can extend the concept to consider it from a commercial perspective as well. The principles that determine how the operating point moves are the same no matter which failure boundary we are concerned with.

We know that we can consider sociotechnical systems to have an operating point that can move based on pressures from management and pressures from the workforce. This movement normally drives it toward a failure boundary. Disturbances also influence the position of the operating point in relation to these boundaries. The force of a disturbance within the system may push the operating point toward a failure boundary. The force of a disturbance from outside the system may actually move the boundary so that it is now closer to the operating point.

The normal movement of the operating point also gives us clues to the system's dynamics and, similarly, to what Cynefin domain it is likely to be operating in. The range of the movement of this operating point within the three system boundaries may be tightly constrained in the case of sociotechnical systems which normally reside in the Clear or Complicated domain of the Cynefin framework. Those sociotechnical systems in the Complex domain are likely to have an operating point that moves within a much wider space within these three boundaries under normal operating conditions. If this area of movement is growing or contracting, something is changing in the system. If it is growing for reasons that are unclear, it may be that the system is going through a domain shift, and so this needs to be reflected in the decision-making mode of the operation (specifically, the SYS4DECIDE mode). In the context of this model, RE is about two things:

1. Keeping the operating point away from the failure boundaries (knowing the location of the operating point).
2. Ensuring that we know the extent of the normal movement of our system's operating point within the area circumscribed by these boundaries and being able to respond appropriately if this changes (knowing the extent of the movement of the operating point, i.e., the area within the boundaries where the operating point has been over a given period).

3.4.4.1 System dynamics at the boundaries

To understand our system's potential behavior when faced with disturbances, we need to understand the dynamics of the system as its operating point is moving toward a failure boundary. There is a good chance that no one recognizes that the operating point is moving in this direction as it may happen slowly or during a period of such increased activity that it goes unnoticed.

How the system reacts as its operating point drifts or is pushed toward a failure boundary can tell us a lot about what we need to do to enhance resilience. Think of resilience as the counterforce that stops the operating point from moving toward a failure boundary across which we run the risk of system failure. The pressures driving the operating point in this direction may be quite intense. The system may be in a situation where demands from the workforce are being met with counterdemands from management, and the vector sum of both pressures will create a strong force that pushes the operating point toward the boundary of unacceptable performance.

For the boundary of unacceptable performance, we need to remember that we do, in fact, have two boundaries to consider. We have the prescribed boundary of unacceptable performance, which is the limit of performance deemed acceptable by the Regulatory Body of the system. We also have the potentially invisible boundary of *actual* unacceptable performance, beyond which we might experience a system failure.

It may also be that the demands to make the system as completely safe and reliable as possible (at a cost) combined with growing pressure for better working conditions from the workforce will drive the operating point toward the economic failure boundary. Beyond this boundary, the organization fails commercially due to spiraling costs.

In our initial description of the Dynamic Safety Model, we showed how the operating point can "flirt" (to use Richard Cook's term) with the prescribed boundary of unacceptable performance and creep further and further into the margin that lies between the actual and prescribed boundaries.[3] It may be that the system becomes so used to having an operating point that lies in this marginal space between the actual and prescribed boundaries that we simply come to accept this as the normal location of the operating point. This is clearly something we need to avoid as once we are in this margin, we have no idea how close we actually are to the actual boundary of unacceptable performance; a slight nudge to the left and the system could collapse.

As we saw in Part 2, feedback loops can work to our advantage, or they can be a problem. Negative feedback loops are the basis of homeostasis. Positive feedback loops are seen in biological reactions to some sort of insult, e.g., the self-creating, self-building blood clotting process that limits blood loss after an injury or the immune system response to an infection.

However, these feedback loops can also be pathological, particularly positive feedback loops. When positive feedback occurs in situations where it isn't warranted, the risk is that this automatic feedback loop will amplify whatever disturbance initiated it until it reaches some terminal endpoint, such as the consumption of all the resources available to it. Positive feedback loops are particularly problematic in sociotechnical systems as they can quickly amplify a disturbance and accelerate the operating point toward a failure boundary. Some positive feedback loops occur spontaneously simply through the nature of the system's dynamics. Other positive feedback loops are initiated intentionally and then maintained despite the problems they are causing. For example, poor commercial performance leads to cost-cutting, leading to worse commercial performance, which leads to more cost-cutting, etc.

Positive feedback in the other direction, such as investment leading to better performance leading to more investment, etc., can be extremely good for the system, but, in general, positive feedback loops need to be tightly controlled. If not, they can rapidly consume resources and drive the operating point in a direction that you might not want it to go.

Before we can take practical measures to improve our system's resilience, we need to understand how the system would behave if the operating point approached a failure boundary, especially the boundary of unacceptable performance. We don't necessarily have to wait for something to go wrong to get this data. It may be possible to model this behavior using stress tests and risk analysis techniques. Daniel Kahneman emphasizes the practice of performing "premortems", the idea being that we imagine that our system has failed and then ask ourselves what the most likely reasons that would have led to this failure are.[4] We can run simulated exercises to see how our system would react in certain situations and, if they are performed honestly, they can give an insight into how our system would behave if it were increasingly being pressured toward the boundary of unacceptable performance.

The most useful source of this data will be the analysis of real-world events that affected the system and shifted its operating point. To reemphasize the point we raised earlier, it is not just the cases where the system got close to, or even passed, a failure boundary that are useful to us. It is also the cases where the system reacted appropriately to prevent too much movement of the operating point toward a failure boundary. Successful outcomes are as valuable a source of system insight as analysis of unsuccessful situations, and this can be exploited if we have a robust connection between SYS4LEARN and SYS4KNOWLEDGE.

Early work into the behavior of systems at the boundary suggested that four factors needed to be considered[1]:

1. Resilience versus brittleness – how successfully and quickly can the system deploy the strategies it has developed to maintain resilience?
2. Buffering capacity – how much adaptive capacity do the components and subsystems have to absorb disturbances and prevent them from pushing the operating point toward the boundary of failure?
3. Margin – how close is the system operating, at any one time, to either the prescribed failure boundary or, if we know where it is (which is far from certain), the actual boundary of unacceptable performance or from other failure boundaries?
4. Tolerance – this is perhaps the most significant when we consider the movement of the operating point toward a failure boundary. Tolerance is about how the system will react when its operating point is close to a failure boundary. Will it degrade slowly or collapse suddenly? Can we expect any warning signs before the system fails? If we can accurately model this, we potentially have a set of signals which we can look out for that will alert the Regulatory Body that the operating point is moving toward a failure boundary.

3.4.5 Systemic failure modes and counterforces

Let's stick with the idea that we aim to ensure that the operating point of our system doesn't cross any of the failure boundaries. To do that, we need to imagine the different ways that it could cross those failure boundaries:

1. Disturbances push the operating point toward a boundary.
2. Disturbances push the boundary toward the operating point.
3. The range of "normal" movement of the operating point expands to the extent that part of it crosses a boundary (suggesting that at any time, the operating point itself could be outside the boundary).
4. Failure of allostatic control to prevent the negative effects of disturbances. As we saw in Part 2, this could be because of repeated disturbances not allowing time for recovery, a failure to adapt, inadequate adaptation, or an inadequate response.

It may be one of these failure modes, or it could be all four of them happening at once, which, in the case of a significant disturbance, could certainly happen. For the most part, when we talk about the failure boundary, it is the boundary of unacceptable performance that is most significant. The economic and workforce boundaries are not insignificant, but the effects of crossing them are more readily understandable (business failure from unprofitability or because of staffing issues).

In the last chapter, we talked about some of the failure modes that can happen on a systemic level rather than at the component level. Most of these are derived from the work of David Woods, and some have been added based on the concepts we have previously covered[5]:

1. Decompensation – a system decompensates when all its global adaptive capacity is used up. Up until that point, it may have been difficult for an external observer to see that the system was compensating at all, but the process of decompensation can be very rapid.
2. Working at cross purposes – when adaptive capacity is deployed in an uncoordinated fashion. This occurs when different parts of the system turn out to be accidently working against each other and so canceling out each other's efforts. Without coordination, subsystems can start seeking their own goals, and the overall system becomes characterized by this multiple goal-seeking behavior.
3. Continuing outmoded behaviors – this can be a feature when centralized control is used in situations that would be better dealt with by decentralizing control. When decision-making is slow or the situation is particularly dynamic, by the time a decision is made, the situation has changed. Instead of revisiting the problem and trying to come up with a different response (or, indeed, working out that a domain shift has happened, and a completely different style of decision-making is necessary), some systems will continue with the old plan while the situation worsens around them.
4. Procedural drift and normalization of deviance – as discussed previously, this is where the system's operating point gradually drifts toward the failure boundary, and the new operating point is accepted as the norm. We can contrast this with the often useful occurrence of efficiency/thoroughness trade-offs that allow the system to continue functioning even when faced with too many procedures and too little time to carry them out.
5. Unrecognized domain shift – failing to recognize that disturbances are pushing the system into a new Cynefin domain requiring a new management style.
6. Boundary encroachment – external pressures may drive boundaries inward. For example, a more challenging commercial environment may mean that the economic failure boundary is pushed inward toward the operating point. We need to consider if the operating point moves to maintain its distance from the boundary or does the gap between them reduce.

With these in mind, we can start to discuss how we can engineer resilience into our system and set up the counterforces needed to stop these failure modes from causing problems in our organization.

3.4.6 Engineering resilience

Now that we've modeled what we think will happen to the system when pressure is exerted upon it by external or internal disturbance, we can start thinking about the measures we can take to build more resilience in our system so that we can improve buffering capacity, expand margin, and identify the signs that the operating point is drifting toward a failure boundary.

Engineering resilience into a system requires working across different scales and across different operating conditions. The features that will make a system resilient need to exist within the system before the time comes when they need to be deployed. As is the way with early attempts to address to such problems, the RE community hasn't reached a consensus about what is or is not good practice. Rather, ideas are developed, which then cross-pollinate with other approaches to generate still newer ideas. In trying to present some practical ways of engineering resilience into systems, I've taken a hierarchical approach in the order in which I present the material:

1. The cornerstones of RE – the high-level features that need to be built into a system for it to have any chance of being resilient.
2. The principles of RE – there are more focused perspectives on issues that particularly affect complex sociotechnical systems.
3. Resilient behaviors – these are specific suggestions about how managers can modify their actions to promote and develop resilience within their organization.

3.4.6.1 The cornerstones of Resilience Engineering

David Woods and Erik Hollnagel identified what they call the four cornerstones of resilience[6]:

1. The ability to anticipate
2. The ability to monitor
3. The ability to respond
4. The ability to learn

If you refer back to the rVSM described earlier in this part of the book, you will see that SYS4 and SYS3 are designed to have these four abilities: SYS4SENSE and SYS3 can monitor; SYS4MODEL can anticipate; SYS4-DECIDE can respond; SYS4LEARN and SYS4KNOWLEDGE can learn. These abilities are also the basis for viability in organisms and are necessary for successful homeostatic and allostatic responses.

These four cornerstones need to exist together to allow them to engender resilience in a system. In some textbooks, they are presented as a closed loop with each experience that is encountered being learned from so that they can be anticipated in the future. A successful system will have this learning culture. This draws on the neurobiological principles of learning in humans: different conditions experienced simultaneously will form common links so that those same conditions allow for a more rapid response in the future.

3.4.6.2 The principles of Resilience Engineering

More recent work has built upon these four cornerstones and has led to the development of some broad principles about how to develop resilience within a system and what dynamical characteristics resilient systems exhibit[7,8]:

- Having tangible experience of surprise – Systems with resilience have often had a tangible experience of surprise. They have reflected and learned from their reaction to that surprise, whether positive or negative.
- Operating with unease – complex systems are precarious. They can behave in unexpected ways, and disturbances may cause effects that come as a surprise. Knowing this, managers of such systems operate with a measure of unease and know that previous or recent success does not guarantee that it will continue in the future.
- Encouraging human adaptive capacity – managers of complex systems need to know that the buffering capacity of their system is most likely to come from the people in it. To this end, human initiative is prized and encouraged. Where appropriate, efficiency/thoroughness trade-offs are considered in their own right as a potential way of improving the system rather than considering every trade-off as a deviation from procedure.
- Developing reciprocity – given the nature of extended adaptive capacity, we know that different components/subsystems may need additional resources at different times. By encouraging reciprocity, it builds into the ethos of the organization that there will be times when one part of the system will need to assist another part. However, at some point in the future, that relationship may have to work in the opposite direction.
- Modeling problems at the right scale – in the same way as our view of what Cynefin domain a system is operating in can change depending on how close we zoom in, so too does the apparent nature of a problem change depending on the complexity or simplicity of our model. Managers accustomed to operating at a particular familiar scale are more likely to try and shoehorn problems they encounter into a model that uses that scale rather than considering the broader context of how the system is operating and relating to its external operating environment. Models that are too complex may slow decision-making unnecessarily. Models that are too simple may miss out important features altogether.

- Requisite redundancy – as many of the case studies will demonstrate, the drive to create a lean organization is often fraught with difficulty. It is possible to strip down a complex system to its leanest version, but that system will only be fit to operate in the conditions that it was experiencing at the time when this "streamlining" occurs. Much like the invisible boundary of unacceptable performance, the amount of streamlining that is planned needs to be thought of in terms of the potential future needs of the systems and in how it will affect the overall adaptive capacity when faced with disturbances. As we saw with theoretic minima in technical systems, trimming back the system structure is potentially very difficult for systems that normally operate in the Complex domain, although slightly easier for predominantly technical systems that operate in the Complicated or Clear domain.
- Maintaining and developing diversity – time and again, research has shown that homogeneity in systems, particularly complex systems, is something to be avoided.[9] The potentially disastrous risks of groupthink can be mitigated, and the variety of the system (using Ashby's term) can be improved by encouraging diversity.[10] The design of the rVSM, like the original VSM, is about ensuring that the system has more possible states (more variety) than can exist in the operating environment it is meant to interact with. In cybernetics terms, diversity is a variety amplifier because a diverse group of people will inherently have more variety in their thinking styles than a homogenous group. Under normal circumstances, this advantage may not be evident, but when it comes time to deploy adaptive capacity, it may be that this diversity allows for the formulation of novel and creative strategies for managing disturbances, especially in situations where there is a need to perform using the knowledge-based level of performance.
- Develop the organization's informational subsystems – resilient performance and system-wide adaptive capacity require well-managed informational subsystems. SYS2, SYS4KNOWLEDGE, and SYS4LEARN will be essential when the time comes for rapid, coordinated responses to dynamic, often novel disturbances. How easily is the information going to flow around the system? Can people get to the information they need quickly and reliably?
- Encourage autonomy – sometimes referred to as polycentric control in the context of organizations, autonomy permits quick, flexible responses to disturbances at a local level.
- Decentralize control when necessary – in highly dynamic situations, it may be necessary to take any limits off the scope of normal, autonomous behavior and decentralize control more fully. There is a balance to be struck between decentralizing control completely, which may hinder coordination between subsystems and trying to keep control central, meaning that decision-making is slowed down, and time is lost through communication channels.

3.4.6.3 Resilient behaviors

It would be nice to be able to list every type of behavior that a manager should demonstrate to develop resilience in her system. Unfortunately, the messy realities of complex systems mean that there will be no complete set of responses. Systems are too complex, too diverse, and too unpredictable.

It's true to say that resilience can be built into the structure of a system. As we have seen, the rVSM has the four cornerstones of RE built into its framework. But beyond structure is adaptive capacity, and RE can be seen as a way of harmonizing the adaptive capacity that components/subsystems are deploying so that it protects the whole organization. To revisit the biological example of complex behavior in slime molds, in times of crisis, all the individual parts come together and mount a complex response that is coordinated solely with the meager biological signaling molecules that every single monocellular organism can generate. In some ways, a large part of RE is the trans-systemic coordination of adaptive capacity.

There are, however, more specific behaviors and perspectives that are associated with resilient systems:

- Even the most reliable system can fail unexpectedly. All you can do is manage the probability of this happening.
- If your system hasn't failed yet, this doesn't imply safety. It could fail tomorrow, next week, or next year.
- Complex systems are not intrinsically safe, and humans aren't always to blame when things go wrong.
- Some events don't have a root cause, and sometimes there isn't anyone to blame. This is covered in more detail in the next section.
- When trying to understand something that has gone wrong in your organization, forget about the outcome. Try to imagine yourself in the place of the people involved as the situation was unfolding around them. They had no idea of the outcome at that time, and it is all too easy to go back and say what they "should" have done now that you have the benefit of hindsight. Doing this may feel like a satisfactory step, but it ignores the more important question: why did the people involved act in the way they did?
- Most people working in your organization try and do their best.
- Bad things can happen to anyone.
- Not following procedures isn't necessarily a bad thing. It can be a sign that the procedures are unworkable, too time-consuming, or that there is a better way of doing something. If people are consistently taking shortcuts, consider that they are probably trying to reconcile two opposing goals, and you should adjust the system so that this doesn't have to happen.
- The gap between work-as-imagined (procedures as laid down) and work-as-done (the shortcuts people take to reconcile multiple goals) should be narrowed as much as possible by incorporating the successful strategies

that people are using to manage trade-offs between multiple goals. However, it may be that the gap is because of procedural drift rather than reasonable trade-offs. In that case, you will need to apply some energy to get the operating point back to where you want it to be by reinforcing prescribed operating practices.

- If there is a big gap between work-as-imagined and work-as-done, you probably have no idea what is happening in your organization.
- It isn't always necessary to immediately change things in the face of failure.
- If you are going to change something, consider what you will change and how those changes may introduce new disturbances into your system. Is it worth the risk?
- The easiest thing to change (the human) isn't always the best thing to change. If many people fail to follow a procedure, changing the procedure may be more successful than retraining everyone.
- Learn from success as well as failure.
- Identify what warning signs you should look out for that might indicate problems in your organization. What are the signs of saturation of adaptive capacity and the subsequent compensation and decompensation that may happen?
- Ensure you have systems in place to guarantee a robust, trustworthy just culture (see the next part of this chapter). It will only take one case where someone is dealt with unfairly or inconsistently for your drive toward developing resilience to stall.

3.4.7 Just culture

Just culture may be a new term to some of you. However, it is a commonly used expression in safety-critical industries.[11] Just culture is relevant to RE because even in settings that aren't safety critical, the people working within those organizations need to know that when things go wrong, events are going to be analyzed fairly. I am considering this concept separately because it doesn't fit clearly into any one part of the preceding discussion. Rather, it touches on many different areas both at the level of components and at the level of the system as a whole. Arguably, it would fit best in the section about SYS5GOAL, specifically with reference to the company's ethos, but I think it makes sense to consider it separately here as well.

Aside from the ethical desire to ensure that people in an organization are treated fairly when something goes wrong, a robust Just culture also has another advantage. In the past, it has been easy when something goes wrong in an organization to use the old-fashioned organizational chart to find a person to blame. The problem is that by quickly deciding that an individual is to blame, the people trying to ascertain what happened miss out on the systemic factors that led to the problem in the first place. This is not to suggest that there are no circumstances where one person or a group of people may have had a direct role in how events unfolded. It is just to say that the actions of one person or a group rarely tell the whole story. Years of analysis of events in safety-critical industries have repeatedly shown that accidents, incidents, and near-misses that happen in complex sociotechnical systems rarely have a single root cause, be that a person, machine, or procedure. More often than not, the event resulted from a combination of circumstances and missed opportunities to stop the chain of events, changes to any of which may have meant that the event wouldn't have happened.

Just culture is not just about reassuring people who work within an organization that they will always be treated fairly. More importantly, it is also a way of ensuring that when these problems do occur, the systemic, as well as the component causes, are fully investigated. Without this, the system cannot learn and anticipate better in the future.

One crucial question that should always be asked when analyzing the circumstances surrounding something that went wrong is whether the same thing would have occurred if the people involved were substituted for other people in the organization. This is known as a substitution test. If the answer is yes, the reasons that led to the problem arising are likely to be systemic rather than down to the people involved. Blaming the individual is quicker; learning about the systemic issues is better.

Another important part of Just culture is the avoidance of hindsight bias. This is the difficulty that people who may be investigating an incident face in trying to get themselves into the mindset of the people involved as the incident was unfolding before them. It is very easy to look back on a chain of events and see all the things that went wrong. It is more difficult to put yourself in the mindset of the person involved during each stage of the event, given that they had no idea that the endpoint of this particular chain of events would be unfavorable. Understanding the time pressures, resource constraints, workload, uncertainty, and the number of choices that had to be made at each step can be a real challenge for people trying to understand these events. However, it is only by doing so that they can get a clearer picture of the systemic causes and so can figure out ways to decrease the probability of them happening again in the future.

Chapter 3.5

Assessing the system properties of your organization

Chapter outline

3.5.1 Introduction

If you are reading this, you may well be in a managerial position within an organization. I can imagine some parts of this book have been more intuitive than others, and the challenge with presenting any work in the field of complexity is how to present it cohesively. To that end, here is a set of questions that you can ask to give yourself some ideas about the potential resilience of your organization. They should also help clarify what the practical applications of the concepts covered in this part of the book actually are.

The questions span everything we have covered on structure, function, capability, adaptive capacity, and resilience. If your answers aren't what you are hoping for, it should be evident how best you could improve your organization from the nature of the question. If you feel that your answers aren't clear or if it was a struggle to formulate them, that lack of clarity suggests that whatever aspect of the organization that you were asking yourself about needs to be looked at to see if it exists in any defined manner.

Designing and Managing Complex Systems. https://doi.org/10.1016/B978-0-323-91609-7.00012-3

The questions are normally phrased in terms of you, your department, and the organization as a whole. However, with the idea of recursion in mind, any reference to management in the sense of senior management is just as relevant to departmental management. If you recall the structure of the rVSM, the whole system needs to be a viable system, and all the SYS1 entities need to be viable systems themselves. They too need to know their identity, have a way of making decisions, and of monitoring and communicating with other parts of the system. Viable systems contain viable subsystems. The other systems (SYS2, 3, 4, and 5) aren't viable systems because they are metasystems within the overall system.

3.5.2 Structure

3.5.2.1 SYS1

- Do you think that the people to whom you need to send information about your work are getting the correct impression of what you actually do?
- Is the information or instructions that you receive from them useful to you?
- Do you do things in ways that you think would surprise the people you report to?
- Do you think you do things in ways that would surprise your colleagues?
- Do you think your job would be easier if you had more autonomy while doing it?
- When decisions need to be made quickly, do they have to be referred upward in the management chain, or do you have the flexibility to do what you think is necessary?

3.5.2.2 SYS2

- How well do you know what is going on in other departments whose functions directly impact you?
- If you need information from them, is there an easy way of getting it?
- Do you think other departments would operate better if they had a better idea of the needs and capabilities of your department?
- Suppose you saw a really significant problem on the horizon that was going to require a company-wide response really quickly. What is the highest level of the management chain you could contact with this information? (Alert signaling network)
- Do you ever think really important, operationally significant information gets lost or diluted in transmission? (Alert signaling network and SYS2)

3.5.2.3 SYS3

- If your department experienced a rapid drop in performance, how quickly would the rest of the organization notice? (monitoring)

- If your department is having difficulties, perhaps due to resourcing or time constraints, how quickly, if at all, do other parts of your organization step in to help? (homeostasis)
- When your organization faces difficulties, is it "business as usual," or do you see changes happening that try and maintain a baseline level of performance? (allostasis)

3.5.2.4 SYS4

- Do you think management always has a clear idea of what they are trying to achieve operationally? (SYS4GOAL)
- How well do you think management knows the reality of the external operating environment within which your organization works? (SYS4PERCEIVE)
- Do you think management is good at anticipating problems that might be around the corner? (SYS4MODEL)
- Do you think that the models used to make decisions are oversimplified? (SYS4MODEL)
- Do you think management is flexible in how they solve problems? (SYS4DECIDE)
- No matter how urgent or how big the problem is, do you think management tries and solves it in the same way as any other problem? (SYS4DECIDE)
- Does a lot of time get wasted trying to categorize a problem rather than fixing it? (SYS4DECIDE)
- Have you seen management act too quickly to try and resolve a problem that could have been better resolved after a bit more thought and analysis? (SYS4DECIDE)
- Is it easy to access the knowledge about how to do your job? (SYS4KNOWLEDGE)
- How much of that knowledge came from tacitly from other people around you, and how much came from formal training and established procedures? (SYS4KNOWLEDGE)
- Is there a big gap between what you see people actually doing and what established procedures and training say they should be doing? (SYS4KNOWLEDGE)
- How much does the company suffer when experienced people leave the organization? (SYS4KNOWLEDGE)
- How easy is it to access and use information related to established procedures? (SYS4KNOWLEDGE)
- Is the organization good at reinforcing information or training that it thinks is particularly important (SYS4KNOWLEDGE)?
- Does your company learn from its mistakes? (SYS4LEARN)
- Does your company learn from its successes? (SYS4LEARN)
- How proactively does your company try to find out what is actually happening in each department? (SYS4LEARN and SYS3, specifically the SYS3 active monitoring submode)

3.5.2.5 SYS5

- Can you define in one or two sentences what the purpose of your organization is?
- Do you have a clear idea of the company's values with regard to how it operates in its operating environment and how it relates to its staff?
- Do you feel that the values of the company change frequently?

3.5.2.6 Transducers, complexity amplifiers, and complexity reducers

- Do you think the format of the information you are presented with accurately reflects the real-world environment it is based on?
- Do you find things get lost in translation as information crosses from verbal to digital format?
- When the operating environment becomes more challenging, do things change in your organization, or are you expected to carry on as normal?

3.5.3 Function

- Do you think you have a clear idea of what your job is based on how it was described when you applied for it?
- If so, do you think this definition is adequate and explains what your job really is?
- Do you think every other person and department have a clear idea of their role in the organization?
- Do you feel that there are important things that should be addressed but that no one is in charge of addressing them?
- Do you and/or others have to do things that aren't your job because not doing them would mean the organization would run into difficulties?
- Do you see your company creating new roles with new job descriptions?
- When these roles are created, do you think it is justified?

3.5.4 Capability

- Do you have access to everything you need to do your job?
- Do you have enough time to carry out the functions that are required of you?
- Suppose you are working at nearly full capacity but the workload continues to increase. Is this met with an increase in the resources that are available to you, most especially, the time available?
- Do you feel that you have the oversight that you need?
- Do you have a clear idea of what you need to do at any one time?
- Have you seen workflows within your department or between departments be slowed down unnecessarily because one part of that workflow didn't have the resources they needed to perform their function in the time allowed?

- When you need to work collaboratively with others, do you think these collaborations are well coordinated?
- Do you see company-wide drives toward efficiency without any rebalancing of resources?

3.5.5 Adaptive capacity

- Do you have the flexibility to achieve your work goals as you see fit?
- Do rules and procedures constrain you too much?
- Do you think other people have more flexibility in how they achieve their work goals than you do?
- When you need additional resources, how quickly do you get them?
- When you need more time, how quickly do you get the changes to your schedule that you need?
- If you really feel that you are working at nearly full capacity, is it easy to offload some of the workload onto other people?
- Are you often called in to help out when another person or department finds that their workload is too high?
- How do you feel about doing this? Do you think it is unfair, or do you think this increases the likelihood that they will help you out if you find yourself in a similar position in the future?
- Do you think your colleagues in your department know when your workload is increasing?
- Do you think different departments know when one particular department is having to deal with a higher workload than normal?
- If one person in the department is struggling to complete their tasks, how does this affect the rest of the department?
- How often do you have to take shortcuts to get the job done in the time available?
- What do you think management would think about these shortcuts?
- Do you think that they might prefer not to know how the job gets done, as long as it gets done?
- Have you ever been surprised when you found out how much someone or some department had been struggling?
- Do you think you have ever had to work at your maximum possible capacity?

3.5.6 Resilience

- Do you think that management and the rest of the workforce are frequently at odds with one another?
- Do unexpected things happen in the organization with no clear reason why?

- How well do you think your organization adapts to changing conditions in its external operating environment?
- How close do you think the organization is to failing, either economically or for other reasons?
- Do you think everyone knows how close or far the organization is from the point where something disastrous could happen?
- If people know, do they care?
- Have you seen the organization operate in a way that you think would not have happened in the past?
- Do you think the organization operates better or worse than it has done in the past? Why?
- Can you imagine something going wrong in the organization that would quickly stop it from operating?
- When the organization is under pressure, do people and management try and continue like it is business as usual, or do you see clear changes in how decisions are being made and implemented?
- Do these changes in management/decision-making style happen often?
- Do you think the external operating environment has forced your organization to work in a riskier way than in the past?
- Do you feel that the organization prepares well for potential future problems?
- If something went disastrously wrong, how willing do you think people in your organization would be to do whatever it takes to keep the system operational?
- Have you ever seen people or departments trying to solve a problem but, in the end, it turns out that their efforts were working against each other?
- When trying to deal with crises or problems, do you think your organization can quickly change its approach if it looks like that approach isn't working?
- Do you think your organization has a recovery plan for the day when something happens that stops the organization from operating?
- When something has gone wrong, is there always an effort to try and find a root cause?
- In your organization, is there always someone to blame?
- When something has gone wrong, will there always be a new rule or procedure to try and prevent that from happening again? Does the addition of these new procedures ever make existing procedural workflows more difficult or impossible to carry out?

3.5.7 Conclusion

These are just a few of the probing questions that you could ask about your organization to try and determine whether it has the structure, function, capability, adaptive capacity, and resilience it needs to continue to be viable.

Perhaps the answers to these questions surprised you in a positive way. Despite the layout of the diagram showing the structure of the rVSM, you might find that the layout of your organization is different but that all the parts necessary are still there, i.e., you have the minimum gene set. Remember, given that SYS1 entities are viable recursive systems within themselves, it is possible to turn the rVSM inside out and represent the whole structure completely differently.

What is more important at this stage of the book is for you to decide whether all the relevant elements are present. It may have been that your questioning highlighted that you don't have access to a reliable alert signaling network or that coordination between SYS1 entities is poor. That doesn't mean your system isn't functioning well now, but it might give some ideas for how it could be refined to operate better in the future.

System design is a necessarily iterative process; with every cycle, the structure gets refined further as more information is collected which tells you about the nature of the operating environment. However, the rVSM structure, the guidance about function, capability, adaptive capacity, and resilience might give you some insights into how you can safeguard your system's viability in the future.

In the next part of this book, we are going to look at some case studies of systems that have or haven't been able to operate in a resilient manner because of how they were designed and managed. Each case study will be followed by an analysis of the events from a systems perspective.

Part IV

Case studies

Chapter 4.1

Challenger and Columbia[1,2]

On a cold but bright morning on the January 28, 1986, a large crowd craned their necks to watch Space Shuttle Challenger rocket skyward. This launch had attracted much excitement as it carried Christa McAuliffe, the first civilian to fly into space. McAuliffe, a schoolteacher from New Hampshire, had been the winner of the Teacher in Space Project, a government effort to inspire students and teachers and encourage children to study science, technology, and mathematics. It was also a media coup for the National Aeronautics and Space Administration (NASA).

About a minute after launch, everything was looking fine from a distance. Only a tracking camera had picked up a small plume coming out of the aft connector of the right-hand solid rocket booster (SRB). Ten seconds later, the right SRB began to detach from the space shuttle, causing it to move off its normal axis. The crew detected this movement. The damage to the solid rocket booster worsened until a fireball developed. This, combined with the increasing damage caused by the disintegrating structure of the booster, led to an explosion that destroyed the space shuttle, killing everyone on board.

Just over a week later, a commission ordered by President Ronald Reagan began their investigation into what had happened. Members included Neil Armstrong, Chuck Yeager, and the physicist Richard Feynman. After analyzing the data, it became clear that the disaster had been caused by a failure of one of the O-rings, large rubber seals that ensure that the separate components of the solid rocket boosters fit tightly together. If they did not fit tightly together, the highly volatile fuel they contained would escape and could be ignited by the rocket's exhaust. In a dramatic demonstration, Richard Feynman immersed an O-ring in ice water and then showed that when it had pressure applied to it, it did not reform into its original shape when the pressure was released. The cold temperatures overnight and during the morning had cooled the O-rings on the solid rocket booster down to a temperature at which they could no longer reliably fulfill their function of ensuring a leak-free fit between the sections of the booster.

As it turns out, the O-rings on the booster had been a cause for concern before. Erosion had been noted on these seals after previous missions, and the booster manufacturers were investigating the problem further. They knew that

Designing and Managing Complex Systems. https://doi.org/10.1016/B978-0-323-91609-7.00013-5

cold temperatures might influence the seals, and some of the engineers working on the boosters were alarmed that missions were continuing without this issue being resolved.

Thiokol, the manufacturers of the booster, put together an O-ring task force whose job was to resolve this problem. Unfortunately, this team felt they were not well supported in the endeavor and didn't have access to the resources they needed to develop a viable solution to the O-ring problem. So bad was the situation in the team that one engineer sent a memo to management that started with the word "HELP!," described the difficulties they were having, and ended with the sentence "this is a red flag."

As the task force was having problems, NASA and Thiokol began to think about how much erosion the O-ring could suffer before failure. With this change in thinking, it now came to be accepted that erosion was a necessary effect of booster use and rather than finding a solution, it could be managed by ensuring that the erosion did not exceed certain limits.

As the date of the Challenger launch approached, there had been some pressure at NASA to ensure that it went ahead as scheduled. The evening before the launch, there had been successive conversations between engineers, managers, and Thiokol because the temperature overnight was forecast to be so low. Several engineers at NASA and Thiokol voiced concerns about the launch, and Thiokol eventually recommended that the launch be delayed. However, after a further debate involving senior engineers and management in NASA and Thiokol, Thiokol eventually signed off on the launch.

Allan J. McDonald, head of the team that manufactured the boosters and the man who received the "HELP!" memo, argued that a delay would be safer and wondered how NASA could rationalize a launch when the temperature was low enough potentially to cause significant problems with the seals. It is recorded in the commission report that he said he would not want to have to explain the decision to a Board of Inquiry. He went further and gave two additional reasons that the launch should be delayed: ice on the launch pad damaging the shuttle and high seas, meaning it would be difficult to recover the boosters (which normally separate from the shuttle when their fuel is used up). He was told that it wasn't his concern. Just before midnight, the approval to launch from Thiokol was sent to the Kennedy Space Center.

The following morning, it was noted by a deputy manager in the Shuttle Projects Office that Thiokol had initially recommended not launching but had changed their advice after a conference at their headquarters. He also noted that the approval to launch had been received in writing. A few hours later, the shuttle launched and exploded shortly afterward.

NASA and Thiokol were heavily criticized in the report of the commission. Allan McDonald had testified that engineers at Thiokol had been overruled when they recommended delaying the launch. Despite the findings of the commission, Thiokol demoted him. On finding this out, the commission threatened to prevent Thiokol from receiving any more NASA contracts. Thiokol subsequently promoted McDonald to vice president, and he was given the job of improving the joints on the boosters that had caused the crash.[3]

In 1988, the space shuttle program restarted with the new boosters designed by McDonald. They were used until the end of the shuttle program in 2011.

Fifteen years after restarting the shuttle program, on January 16 2003, Space Shuttle Columbia, the first ever space shuttle to enter service, launched into space. Several important experiments were going to be carried out during the two weeks that it would spend in orbit.

Shortly after the launch, NASA engineers reviewing the video footage noticed something abnormal. As well as two solid rocket boosters, the space shuttle also has an external tank that supplies the shuttle engines with fuel during the launch. This tank normally detaches a few minutes after the solid rocket boosters detach and disintegrates in the atmosphere. The external fuel tank is sprayed with a deep layer of polyurethane foam to insulate the liquid oxygen and liquid hydrogen stored inside. From the looks of the video, it appeared that a large section of foam had broken away from the tank and had hit the shuttle's left wing.

The wings and underside of the shuttle are covered with an extremely heat-resistant material designed to prevent the heat generated during reentry from damaging the shuttle's structure. Engineers were concerned that this large piece of foam could have caused damage to this layer of material.

Much like the problems with the O-rings, there had been a long history of problems with foam breaking away from the outside of the external fuel tank during launches. On multiple occasions, it had been found that foam had damaged a space shuttle to some extent, even if this damage was only detected once the shuttle had landed back on Earth. The accident report stated that early in the space shuttle program, engineers were concerned about the amount of damage this foam could do if a sufficiently large piece broke away and hit the shuttle.

As the space shuttle missions continued, foam strikes became a regular occurrence but did not cause severe damage to any of the shuttles. It began to be accepted as inevitable and unlikely to affect the shuttle's safety. Despite the initial shuttle design parameters stating that the external fuel tank should not shed debris and that the shuttle should not be struck by anything that could impart more than a trivial amount of force, the regular shedding of foam and impact on shuttles being launched was accepted as the norm. The report states that this was an occurrence of the normalization of deviance. This normalization is reflected in the NASA documentation, which initially had foam-shedding listed as a serious concern and then later relisted it as an "accepted risk."

Having seen that some debris had struck the shuttle, a Debris Assessment Team was formed at NASA to assess how this might affect the mission. Several missed opportunities are listed in the accident report, any of which might have given the engineers on the ground a much better idea of the amount of damage that had actually been caused. One particular email that was written but not sent was from a structural engineer. In the email, he emphasized that it was irresponsible not to try and get additional imaging of the impact point and that if any of the heat-resistant tiles had been knocked out, it would represent a grave danger. The email ended with a restatement of NASA's safety motto: "If it's not safe, say so." When asked why he didn't send this email, the engineer in question said that he didn't want to jump the chain of command and would defer to management's judgment on whether to try and get additional images. It had been suggested that the crew perform a spacewalk to assess any damage, but this suggestion was not acted upon.

NASA elected not to explore the problem any further. They informed the crew on the shuttle of the potential impact just in case it was mentioned in an upcoming media interview.

Had the extent of the damage been discovered, the story of this Columbia mission could have been a success story like that of Apollo 13 rather than a disaster story like Challenger. The accident report presented several theoretical repair and rescue scenarios, but none were considered at the time.

On February 1, 2003, Space Shuttle Columbia began its reentry. As the heat began to build, telemetry at NASA showed that four hydraulic fluid temperature sensors in the left wing had stopped working. Shortly afterward, the shuttle lost more of its thermal protection layer. Further indications on the ground showed that the tire pressure sensors on the left landing gear had failed, suggesting that the immense heat generated on reentry was now penetrating through the protective thermal layer into the landing gear bay. From then, a cascade of system failures occurred, and Columbia disintegrated in the skies over Texas.

As part of the subsequent investigation, engineers took a section of the wing from one of the other shuttles and used a compressed air gun to hit it with a block of foam the same size as the one that hit Columbia. It blew a hole 16 inches in diameter through the thermal protection layer.

4.1.1 Analysis

Except for financial markets, it is hard to think of a more tightly coupled, intractable, complex sociotechnical system as the human space flight program. The Challenger case study could have been just an example of a component failure leading to a system failure; the O-ring fails, and the shuttle is destroyed. The commission that investigated the accident was actually very forward-thinking to the extent that they investigated the systemic issues that contributed to this component failure. From a systems perspective, we can identify several issues and some missed opportunities:

- There was immense organizational and political pressure to get more launches carried out. As well as carrying Christa McAuliffe, the Challenger shuttle was also tasked with putting a satellite into orbit. To try and keep on schedule, the economic boundary pushed inward and began to force the system's operating point toward the boundary of unacceptable performance, one of the failure boundaries of the Dynamic Safety Model.
- Concerns about the O-rings had grumbled on for several years. O-ring problems began to be accepted within both NASA and Thiokol despite finding evidence for erosion during previous launches. Rather than stop the shuttle launches until a redesign of the joints could be carried out, this erosion became normalized, and instead of not accepting it, the question became how much erosion could be accepted.
- There was disagreement and mistrust between engineers and management (some of whom were also engineers). With a drive to deliver on time, management pushed back against the concerns of engineers to the extent that the head of the manufacturing program refused to approve the launch. Thiokol approved the launch anyway. If some sort of alert signaling network had been available to make the flight operations director at NASA aware that this was a red flag issue, perhaps it would have been enough to delay the launch.
- The commission report highlighted that the management structure of NASA was such that flight safety problems such as this were able to bypass key managers who would have had the authority to intervene. This would seem to be a problem with SYS2 and SYS3.
- In 2003, the commission that investigated the loss of Space Shuttle Columbia noted that despite the commission's recommendations after the Challenger disaster, NASA still had the same flawed decision-making process and that this had resulted in the destruction of Columbia. The report even highlighted the normalization of deviance with the acceptance of foam strikes during missions in the same way as it accepted O-ring erosion after booster use.
- In both cases, there were many opportunities to intervene. It is clear from the accident reports that at various times during the run-up and during the course of the event, there were engineers with serious enough concerns to try and raise the red flag. In the case of the unsent email, if there had been an alert signaling network, even an anonymous one, the engineer in question may have been more likely to send it.

Chapter 4.2

Walmart, FEMA, and Hurricane Katrina[1–4]

During the latter part of August 2005, a tropical depression built up over the Bahamas. As it moved across the warm waters of the Gulf of Mexico, it rapidly increased in size, and by the time it made landfall in Louisiana, it was classed as a Category 3 hurricane. The heavy rain and the storm surge flooded large parts of New Orleans. This flooding was exacerbated by the failure of the levees designed to prevent such flooding. The storm persisted in the area of New Orleans, and the high winds and rain caused significant damage to houses and infrastructure. Many roads in and out of New Orleans were rendered unusable, and with such damage to transportation infrastructure, the aid efforts were hampered.

The Federal Emergency Management Agency (FEMA) was designed to handle just this sort of crisis. Able to be deployed anywhere in the United States at the order of the President, it is supposed to be able to organize evacuations, undertake search and rescue, and provide temporary housing, supplies, and medical aid.

Unfortunately, in the case of Hurricane Katrina, the federal response did not turn out as planned. With FEMA being headquartered in Washington and several regional offices throughout the country, their response to the disaster was typified by a desire to maintain centralized command and control. Interaction with other government agencies was uncoordinated, and time was taken up in trying to determine what agency was in charge of the situation. Without local knowledge of New Orleans, FEMA wasn't familiar with the general geography and topography of the landscape and seemed not to appreciate reports coming in that suggested that the levees were breaking down and allowing millions of gallons of water to flood nearly 80% of New Orleans. Evacuation efforts were slowed down when FEMA rejected offers of additional buses and boats to aid in rescuing stranded citizens and transporting them to safety. Senior management in FEMA insisted that all rescue and evacuation efforts be coordinated through state agencies using the proper channels going as far as telling emergency services not to respond to calls unless they had been ordered to do so by the correct authority. Trucks carrying supplies of water and diesel were turned away, and FEMA confiscated hospital

Designing and Managing Complex Systems. https://doi.org/10.1016/B978-0-323-91609-7.00014-7

supplies that had been donated. To ensure that urgent medical supplies reached the hospital, the donor, the out-of-state owner of the hospital, shipped a second consignment of supplies to a different part of the state and had them flown directly to the hospital by helicopter.

There is also evidence that the information coming to command-and-control centers from officers in the field was scant, leading to an incomplete understanding of the scale of the situation. For example, it wasn't until three days after the hurricane hit that the FEMA Director found out that there were refugees in the large Convention Center in the city.

FEMA was heavily criticized because of its poor response to the disaster. Its director, a political appointee with limited experience in disaster management, resigned from office shortly after he had been recalled to Washington and been replaced as the lead person responsible for the disaster management efforts.

Various investigations and pieces of research have been carried out over the years looking at the federal response to Hurricane Katrina, particularly concerning disaster management in New Orleans. Among the findings, several comments about FEMA's approach are relevant to what we've covered in this book:

- The person assigned to supervise FEMAs response had limited experience of disaster management.
- Coordination between agencies at the state and local levels was characterized by a general lack of organization and a high index of suspicion.
- The FEMA response was extremely bureaucratic, with the focus being on doing the proper thing rather than the appropriate thing. One research paper noted that FEMA was characterized by an intense desire not to commit an error by doing something too quickly.
- Coordination within the agency was very poor. Supplies were routinely sent to multiple different locations, frequently out-of-state, and stored at the government's expense. In one of the most surprising stories, FEMA ordered 182 million pounds of ice to be delivered to stranded families, but much of this ice ended up in storage facilities all around the country before being melted down.
- In a disaster situation, three different levels of management are involved in the response: management in Washington, management in the region, and management in the response team on the ground. Even to those in FEMA, how these three levels coordinate is unclear, which exacerbates the internal bureaucracy and slows the decision-making process.
- FEMA was clearly underresourced even in the run-up to Hurricane Katrina. Budget cuts had led to cuts in staffing, and they did not have access to reservist staff who could quickly join in disaster relief efforts. The United States Senate report into the management of the aftermath of Hurricane

Katrina also noted that many of FEMA's senior staff were political appointees with little emergency response experience.
- The knowledge base of FEMA had also been eroded as budget cuts led to experienced staff leaving the agency and taking important, practical, institutional knowledge with them.

Fortunately for the people of New Orleans, where the federal government failed, another large, wide-reaching organization demonstrated its remarkable adaptive capacity in the face of this incredible challenge.

Walmart is a huge multinational retail company with general retail and grocery stores in countries worldwide. Based in the United States, they had developed the necessary business continuity plans that could be put into action in the event of natural disasters or other business-critical events. As they own so many stores throughout the world, they have an emergency command center at their headquarters in Arkansas which is staffed normally by 6–10 people. So that it can respond to emergencies of different scales, the facilities and staffing of the command center can be expanded and can include representatives from other departments in the company.

With its own hurricane tracking systems, Walmart was already massing supplies at staging areas just outside the hurricane's expected path with drivers in place to bring them into the disaster zone as soon as it was safe to do so.

Although the reason behind this careful planning and preparation was primarily business continuity, i.e., getting damaged stores repaired and restocked as quickly as possible and supporting staff on the ground, the damage caused by Katrina was so extensive that Walmart's response went way beyond that.

As it became clear that the damage to stores, housing, and infrastructure was likely to be severe and that the situation on the ground would be extremely dynamic in the immediate aftermath of the hurricane, Walmart did one crucial thing that enhanced their responsiveness well beyond anything that FEMA would be able to manage: they decentralized control. Walmart CEO Lee Scott passed a message down to the regional, district, and store managers: "A lot of you are going to have to make decisions above your level. Make the best decision you can with the information that's available to you, and, above all, do the right thing."

This permission for employees to deploy their adaptive capacity was what characterized the successful response of Walmart in supporting the disaster relief and recovery effort:

- An assistant manager used a bulldozer to break into the ruins of her store to salvage what she could. She then left this in the parking lot for residents to take for free. She also broke into the pharmacy and brought all the drugs to the local hospital.
- At another store, an employee broke open the warehouse door with a forklift to bring badly needed drinking water to a local retirement home.

- Satellite links were set up with stores so that they could communicate with headquarters.
- Stores that were too badly damaged came up with improvised tented areas in the parking lot to be able to provide residents with some basic necessities.
- Trucks that had been prepared in advance were able to deliver supplies into the disaster zone rapidly. In the three weeks after the hurricane, approximately 2500 trucks with supplies made it to New Orleans.
- Walmart coordinated with other organizations willing to donate supplies and used their trucks to bring them into New Orleans. They were also able to help provide supplies to people who had been evacuated out of Louisiana to other states.
- They announced that they would fill prescriptions free of charge to residents and evacuees even if they didn't have a copy of their prescriptions.
- Recognizing the importance of maintaining communications, Walmart distributed generators and communication systems to help with coordination and also to allow people to locate relatives.
- Significantly, the company also recognized that for business continuity, there had to be a community there that needed them. This perspective is important because it showed insight when it came to modeling the future status of the Walmart system in relation to its operating environment. If their relief efforts had failed and communities had been wiped out or displaced, Walmart would have had no communities to supply their customers. This is not to say that this was the main reason behind their efforts, but rather it shows that there is a potential symbiosis that exists between commercial systems and the environment in which they operate.
- One of the researchers that investigated the success of Walmart in dealing with the aftermath of the hurricane noted the powerful corporate culture (ethos) in Walmart and mentions the beliefs that are part of the company makeup: respect for the individual; service to the customer; striving for excellence; acting with integrity.

4.2.1 Analysis

The contrast between the FEMA and the Walmart responses is based on a willingness to decentralize control but also on several other considerations that we have looked at in this book:

- The hurricane caused a domain shift. The system of New Orleans was in the Chaotic domain, a domain where the appropriate management strategy is act-sense-respond. Walmart's willingness to decentralize control gave employees on the ground the ability to act immediately to try and stabilize the situation. When they had done this, they could start to make further assessments as to what their needs would be and communicate this back to headquarters, who could then respond by deploying the necessary resources. FEMA remained rooted in Complicated domain thinking and went to great lengths to analyze what they thought the problems were. They also spent an excessive amount of time trying to reconcile rules from across various agencies to try and guarantee that there would be proper coordination between agencies. All this came at the expense of the immediate action that could have helped the situation.

- From reading accounts of the event, it seems that Walmart's sense of urgency in setting up a communication network between stores and headquarters was not an effort to reassert central control but rather to provide a channel to help coordinate efforts locally. This is a great example of SYS2 and SYS3 being used to coordinate local adaptive capacity and ensure that autonomy doesn't lead to working at cross purposes.
- The whole FEMA structure appeared not to be viable. Given its function, it is clear that budget cuts had eroded the structural elements required to mount an appropriate response to the extent that they may have actually worsened the situation. The loss of staff meant that SYS4KNOWLEDGE had been depleted, and the political appointees that were in charge of the Regulatory Body didn't have the necessary knowledge and experience to plug this gap.

Given that commercial organizations such as Walmart necessarily keep large stocks of resources that would be needed in such an emergency, there has been some speculation that private enterprises may be in a better position to handle disaster relief of this type compared to government agencies.

Chapter 4.3

Lake Peigneur[1,2]

In 1980, the Texaco mining company was given permission to perform some exploratory drills in Lake Peigneur, a large freshwater lake in Louisiana in the United States. They had high hopes that there would be oil in the area because it was known that there were large, subterranean salt deposits, and it was common to find both oil and salt deposits in the same geographical area.

Lake Peigneur was a popular lake for fishing and water sports and boasted a large nature reserve and botanical garden. Texaco planned their drill sites and proceeded with their exploratory operation.

With the help of several barges, a drilling platform was positioned at a certain point in the 10-foot-deep lake, and drilling commenced. At approximately 1300 feet deep, the drill bit became stuck. Soon, the rig began to tilt, and the crew evacuated the site.

Over the next few hours, a massive whirlpool formed at the site where the drilling occurred and dragged the massive oil rig, its stabilizing barges, and billions of gallons of water into an ever-widening hole. The oil rig was 150 feet high, and somehow it had sunk into a lake that was meant to have a depth of 10 feet.

So great was the suction force that the land surrounding the lake was sucked downward too, along with boats, trees, and houses. A canal that normally led out of the lake and flowed into the Gulf of Mexico reversed its flow and turned into a waterfall that was 150 feet high as saltwater flowed backward to fill the lake.

It was unfortunate for Texaco that another large sociotechnical was trying to take advantage of the natural resources in the area. The Diamond Crystal Salt Company had a mine on Jefferson Island that had been established many years earlier. Below the lake was an extensive salt reserve, and vertical and horizontal shafts had been dug throughout this reserve to mine out the salt. The location of this salt dome was well documented on geological charts, but it appears that in calculating the position of their exploratory drilling site, an engineer had used a coordinate system that was incompatible with the planning procedures. This resulted in the shaft being dug too close to the mining operation.

As the disaster made it difficult to assess what actually happened, it is likely that the drill approached close to or actually pierced one of the horizontal tunnels within the salt deposit. The pressure of the water from above and the subsequent subsidence of the earth as the mining tunnel collapsed

Designing and Managing Complex Systems. https://doi.org/10.1016/B978-0-323-91609-7.00015-9

widened the hole bored by the drill to a sufficient degree to allow the lake to flow out through the mine. Fortunately, all the miners could escape, and there were no fatalities above ground.

Texaco eventually paid compensation to the mine owners and the local businesses who had suffered because of this disaster. Lake Peigneur has gone from being a 10 feet deep freshwater lake to a 200 feet deep saltwater lake.

4.3.1 Analysis

The reports available on this accident do not delve deeply into organizational decision-making in the same way as the reports of the space shuttle crashes do. Because of this, it is difficult to draw conclusions about what elements of the system could have been changed to mitigate the chances of a disaster like this happening.

Several of the other case studies in this section look at what Charles Perrow would call "system accidents." Lake Peigneur is something a bit different. We could say that there is certainly an element of component failure here, i.e., the incorrect coordinates being used which led to the drill site being in a different place than it was meant to be. On top of that, there are likely to be some systemic issues, i.e., not confirming the siting of the exploratory platform given the knowledge that there was salt mining going on in the area. Additionally, there is another phenomenon here that is worth looking at.

The oil mining operation and the salt mining operation are individual systems in their own right. Neither is dependent on the other, but because of the quirks of geology, where salt is found, oil is likely to be found, and so these systems operate in the same block of physical space. If the drilling site had been correctly located, it is likely that these systems would never have had any reason to interact with each other.

The phenomenon here could be referred to as intersystemic failure. We have been mostly looking at failures that can occur in single systems because of internal or external disturbances. Here, we have a case of two systems operating normally that suddenly find themselves occupying the same physical space as each other.

In Part 2, we saw how it is normal for systems to interlock in a planned way, e.g., the output of one productive system will be the feedstock for another productive system. These intersystemic interactions can be managed. What is more difficult to manage is when separate systems intersect unexpectedly. The reaction of both may be to deploy their adaptive capacity to deal with the disturbance, but as these systems are colliding, it may be that the adaptive capacity of one is working counter to the adaptive capacity of the other.

This type of intersystemic failure has not been widely researched, but, as in the case of Lake Peigneur, it may be that these types of failures are potentially the most damaging.

The phenomenon in intersystemic interaction is more important to consider at the planning stage. In designing a system, it is usually clear what other systems will have to be interacted with. A trucking company knows that it will have to interact with a company that provides fuel in the same way as a train company will need to interact with the electricity grid. These interactions are predictable and necessary. Even when they are, there needs to be some procedural organization between the two systems that defines how they can interact. For example, a fuel truck arriving at an aircraft might bear a sign saying only fuel truck employees can access the lifting platform that allows the hose to be connected to the aircraft's fueling panel. The airline might also have a rule that says that only airline staff can operate the fueling panel. To get the job done, some sort of intersystemic trade-off is going to have to happen.

The difficulty with intersystemic trade-offs is that they are unconstrained.[3] Within a system, we know that the operating point moves toward failure boundaries depending on the disturbances that are happening inside or outside of the system. An efficiency/thoroughness trade-off can mean that the operating point shifts, but, hopefully, we have designed enough resilience into our system to detect this, and we can employ adaptive capacity to push the operating point away from a failure boundary. But what happens when trade-offs happen because of conflicting requirements between two different systems? Which system is responsible for detecting whether this disturbance occurring between systems rather than within one system will have an effect? Coordination within a system is clearly very important, but coordination between systems is even more important simply because the effects of uncoordinated interaction can be so unpredictable and so damaging.

Other interactions with other systems can be significant even if the systems themselves aren't interdependent. Two systems trying to achieve two different things within a defined block of space and at the same time may find that their goals are not compatible. When two systems are occupying the same space, they are essentially overlapping, and the whole combination is made up of subsystems that are exhibiting multiple goal-seeking behavior in an unconstrained and uncoordinated way.

Although it was not intentional, the unexpected interaction of the mining company and the drilling company within the same physical space simultaneously demonstrates the problem very clearly. When the same thing happens between systems that are normally separated, how they react when they encounter each other for the first time may be what defines how badly things are going to go.

Chapter 4.4

The water temples of Bali[1–3]

In 2012, the United Nations Educational, Scientific and Cultural Organization (UNESCO) designated the Balinese Water Temple system as a protected cultural landscape and a manifestation of the Tri Hita Karana philosophy that reflects the combination of the spiritual world, the human world, and nature.

Bali is an island of densely terraced rice fields. Over more than a millennium, the rice farmers have developed a unique system based around collective groups of farms. These groups are called subaks.

Being a volcanic island, rainwater flowing down the sides of mountains is enriched with minerals, some of which act as natural fertilizers. Dense terraces suitable for planting have been built into the hillsides over centuries to use this rainwater most efficiently. Given that the water flows downward toward the sea and these terraces exist at different heights above sea level, subaks are designed to help divide water resources fairly among groups of farmers. Using traditional building techniques and materials, channels cut along the mountainsides collect rainwater and divert it into a network of channels that run between and into the groups of farms so that their crops can be irrigated.

To ensure that farms, no matter their height, i.e., their proximity to the water source, receive a fair share of the water, over the last millennium, groups of farms have congregated together to form subaks. Each subak receives a water channel from the main canal which flows via a water temple. The water temple acts as a cultural and physical link between the farms forming the subak.

So that collective decisions about water usage are made fairly, the subak system also includes a democratic management structure. Farmers in a subak come together and elect a leader who serves for a fixed period. During subak meetings, farmers collectively plan their planting, crop rotation, harvest schedules, as well as determining which religious rites should be performed at different points in the growth and harvesting cycles. Most importantly, they also decide how water flowing into the subak is divided between the different farms.

The subak members allocate resources based on availability. If water is plentiful, all farmers within the subak can grow rice. If water is restricted, the farmers arrange to rotate crop growing in the different farm areas to include crops with a lower water requirement. By carefully planning crop rotations, the subaks also reduce the risk of pests and plant disease by disrupting the life cycles of these destructive organisms.

Designing and Managing Complex Systems. https://doi.org/10.1016/B978-0-323-91609-7.00016-0

Despite Balinese society being culturally stratified, the subak system is designed to ensure that all farmers participating within it can do so on the basis of equality irrespective of their background or the size or location of their farm. During subak meetings, formal pronouns usually used linguistically to denote status are not used to ensure that every member is treated equally. Through this bottom-up management structure, the subak system has produced the optimum arrangement of farms and planting cycles as confirmed by computer simulations. Efforts by the Balinese government to try and centralize control over the subaks led to results that were inferior to the system of organization that subaks had created for themselves.

In the 1970s, the supposed Green Revolution tried to increase rice yields from the subaks by promoting intensive agricultural techniques. However, because of the high planting density throughout the year, the pest life cycle reasserted itself as there would be food to eat all year round. To eliminate this problem, subak farmers were ordered to use pesticides, which failed to address the problem. It also ran counter to the Tri Hita Karana philosophy, which respects all living creatures, even pests.

Eventually, the subaks were allowed to return to their own form of management which has led to remarkable crop yields given that chemical fertilizers and pesticides are not used.

In a resource-limited situation such as this, you might wonder why the farmers that have farms located in more naturally advantageous locations, i.e., close to the source of irrigating water, don't capitalize on this. It would mean that they would always be able to plant rice rather than a less profitable crop. Perhaps competition rather than cooperation would be a better option.

Such has been the experience in the past that farmers know that this type of approach is worse for everyone. If a farm higher up uses most of the water for its planting season, farms lower down will have to wait until the farm higher up doesn't need such high quantities of water so that they can use it to irrigate their plants. Rather than synchronized planting, neighboring farms now plant rice at different times, and this means that the pests which damage the rice simply have to migrate from farm to farm throughout the year. By sharing the water to allow synchronous planting, the pests die out.

Although the physical and organizational arrangements of subaks are remarkable in themselves, the final element of the Tri Hita Karana philosophy is that of the spiritual world. Through the water temples, the spiritual lives of the farmers are also interconnected, fostering a deeper will to cooperate rather than compete. In a country dependent on rainfall, freshwater flowing down from the mountain is considered holy. Religious ceremonies on the island often involve rice, and ceremonies are designed around prayers for a good harvest or prayers of thanks after the harvest has been collected.

High up in the mountains on the edge of a lake, a natural reservoir for the water running off the mountains, is Pura Ulun Dana Batur, the Supreme Water Temple. Every year, a representative from each subak travels to the temple with an offering. The temple priests receive the offering and give each delegate a bamboo vessel containing water from the lake which has been blessed. This water is then brought to the water temple of each subak where it is mixed with local water that has also been blessed. The farmers and their families drink some of this water to share the blessing, and every field also receives a blessing using this water.

While there is organizational alignment within the subak system, the rituals surrounding the water temples also promote spiritual alignment between the members. Rather than competition, the Tri Hita Karana philosophy promotes harmony between the gods, the humans, and with nature.

4.4.1 Analysis

The subak system allows us to look at system interactions at a different scale. Under normal circumstances, a Western perspective would probably consider a single farm to be a sociotechnical system in itself: humans, machines, and procedures (planting times, milking schedules, etc.) that come together to grow crops and raise animals for food.

Individual farms within the subak system have the same structure and function. However, to maximize productivity in the face of a limited resource (water), the subaks naturally evolved (one might say their system was autocatalytic in how it created itself) so that over longer periods, the entire network of farms was maximally efficient.

In the example of Lake Peigneur, we looked at what happens when two systems collide in an unexpected and uncoordinated way. Similarly, if individual farms wanted to monopolize resources, the process would have been ultimately self-destructive when the pest life cycle reasserted itself. Long-term losses would cancel out short-term gains.

When distinct systems decide to benefit by cooperating, they essentially agree to become subsystems within a bigger system. They autocatalyze themselves into SYS1 entities which then necessarily need to have the SYS2, 3, 4, and 5 metasystems that the management and religious parts of the subak arrangement provide. Here, a viable system has created itself at the SYS1 level and has created the coordinating metasystems that allow it to thrive.

Chapter 4.5

The global financial crisis[1–3]

This is the case study that I have been dreading writing. Even in the 1980s, Charles Perrow recognized that the international financial system was the most highly coupled and least tractable system that humans had ever created. Since then, the technological explosion in investing infrastructure has only made the whole system more opaque and more tightly coupled. Under normal circumstances, we could say that it operates in the Complex domain of the Cynefin framework, but during 2007 and 2008, a critical domain shift occurred that drove it into the Chaotic domain. It is a complex story of positive feedback cycles and a system that was created without a Regulatory Body (SYS3, 5, and 5). So many things went wrong in such a chaotic way that on August 9, 2007, the system, such that it was, essentially broke apart. Going into the crisis in detail would fill many books, so we are just going to pick out some elements that we can put in the context of what we have covered on system design and management. Even with the narrow focus on one part of the crisis, I've simplified some of the actual events, but the principles are still correct.

Before the crisis started, financial institutions around the world, whether private or governmental, were like SYS1 entities in the broader system that is the global financial system. You might ask why we can't just consider these banks and other financial institutions as systems in their own right. To a certain extent, we can: banks will have a corporate structure and engage in commerce with other entities in their operating environment. A lot of the time, those entities may be normal customers or small businesses. Banks will also interact with large businesses, and most importantly for the purposes of this case study, they interact with each other.

We know from Part 3 that any SYS1 entity must be a viable system in itself, so to understand the global financial system, we must consider all these separately viable banks as loosely linked within a system. The reason that we, as time-traveling observers looking at this arrangement, need to take this perspective is that the global financial crisis was caused by the mismanagement of debt between these institutions. This is how a housing bubble in the United States nearly led to the collapse of the global financial system.

Mortgages are normally a secure way for banks to make money. People want to buy houses, so banks give them the money to do so and expect it to be paid back over several years with interest.

Banks also like to make money by making deals of their own. To make a potentially good deal even better, banks can borrow money from other

Designing and Managing Complex Systems. https://doi.org/10.1016/B978-0-323-91609-7.00017-2

institutions so that they have even more money to invest. If the deal goes well, the bank can make a lot more money than if it only had its own cash to invest. Now they get to keep all the profits, but they must pay back the money they borrowed from the other institution plus some interest. This method of investing with borrowed money is called leverage, and leverage allows people to make a lot of money by taking out loans to invest even more in a particular deal. The big risk with leverage is that if the deal goes badly, they can lose so much money that they won't be able to repay their loan.

Because of this, when getting a loan, most institutions will require some sort of collateral. They know that sometimes deals go badly, and if the person they are lending to can't pay back the money, the institution can seize whatever collateral the borrower used to support their loan. For most normal bank customers, we use the house that we buy as collateral against the loan that we borrow from the bank. That loan that we borrowed means we can make a deal that requires a lot more money than we have. We are leveraged, and if we don't pay back the loan, the bank will seize our collateral, i.e., repossess our house. Getting a mortgage is a simple example of collateral and leverage. What happened in the years running up to 2008 demonstrated that positive feedback without a Regulatory Body can bring huge systems crashing down.

A bank with lots of customers who have mortgages knows that these customers (or most of these customers) will continue paying their mortgage because they don't want the bank to seize their collateral (their house). Banks often need money to make deals of their own, and so to get more cash, they take all the mortgages and create a financial product called a Collateralized Debt Obligation (CDO). Because CDOs pay a regular income and are backed up by lots of collateral (people's houses), other large financial institutions are happy to buy them, and so the bank gets cash, and the other financial institution gets a product that continues to pay them a steady income. So far, so good.

Here's where things started to go wrong. The financial institution that owned the CDO now wants to make some deals of its own. It wants to borrow money from some other institution to do so. That institution might be happy to give them the cash, but in the same way as with the mortgage, they will want that cash repaid with interest, AND they will want some collateral just in case the deal goes badly. The first institution uses the CDO that they own as collateral.

Some other institutions holding CDOs decide to split those CDOs and use the pieces to create still more CDOs (a "CDO-squared"). All these steps involved leveraging up the amount of cash involved in the deals so that with so many loans backed by CDO collateral and CDOs being leveraged up into other CDO investment vehicles, what started as a CDO that was backed by, for example, $100 million of mortgage loans now became a monster that might involve different institutions owing other institutions over $1 billion.

The CDO industry was spectacularly successful at making money. Every step of the process of creating and selling CDOs involves fees to the people managing this process. Therefore they were keen to create these mutant CDOs in as many forms as possible. The problem they encountered was that they needed a big batch of mortgages for each mutant family of CDOs to act as the first seed out of which this monster plant would grow. However, banks were limited in how many mortgages they could offer because there were only so many people in the country who had the financial security to afford mortgages.

Given that house prices were rising year-on-year, banks and other institutions thought that they could start taking some more risks and giving mortgages to people who ordinarily wouldn't have the financial security to make them eligible to get one (e.g., a job that was paying enough or evidence of other assets). With house prices rising, they figured that if the worst came to the worst and people stopped paying their mortgages, they would be able to seize their houses, which would be worth more than when their clients had bought them. To hook these people into getting a mortgage (and creating more material to power the booming CDO industry), banks would offer low-interest rates that would last for a couple of years before the normal, higher interest rates kicked in.

Tragically, two positive feedback cycles clashed after years of running this process. The first was the upward spiraling positive feedback loop happening in the CDO market: CDOs were enabling more leverage, enabling more deals that enabled more CDOs that were enabling more leverage, and so on. The second positive feedback loop started when all those mortgage loans that were made to people who ordinarily wouldn't have been able to afford them then started switching from the low-interest rate that had been offered for the first couple of years to the normal, higher interest rate that the bank would charge from then on. Suddenly, people's monthly mortgage payments increased dramatically, and many couldn't pay the higher rate. When they defaulted on their mortgages, the banks seized their houses and put them up for sale. Because this was happening all over the country, there was suddenly a huge increase in the number of houses for sale, and this caused the housing market to drop. Suddenly, people who had these mortgages and had expected the value of their houses to keep rising discovered that their homes were worth a lot less than when they had bought them. Mortgage defaults continued to happen, and banks were left with repossessed properties they were trying to sell in a falling market. Although this looks like it's going in a negative direction (which it certainly was), this is still technically a positive feedback loop: more financial problems, more defaults, more repossessions, more houses on the market, lowering prices, more financial problems, and so on.

For that first bank that sold the first CDO in the example we talked about, they didn't get into too much trouble. They had taken the mortgages and sold them for cash. It was the financial institution that bought the CDO that started having trouble. First, the CDO stopped providing the monthly income that should have come from people paying their mortgages. Secondly, if they wanted to get their money back by seizing the collateral (the houses), they found that those houses were worth a lot less than they had been in the past. If that wasn't bad enough, all the leveraged deals built on the back of these mortgages now started to unravel. At each stage of unraveling, the situation worsened because of the ever-increasing amounts of leverage involved in the deals. The collapse of the various chains of CDOs because of the failure of the initial mortgages amplified the amount of money lost into many times the value of those initial mortgages. The positive feedback loop spiraled upward until nearly all the money tied up in these esoteric CDO-backed deals vanished.

With such large amounts of money disappearing from the balance sheets of the institutions that owned them, the institutions themselves began to disintegrate. With the global financial system depending on banks lending to other banks, all of it ground to a halt. With so much money disappearing from their balance sheets, banks stopped lending money to anyone, especially other banks, so that they could keep whatever cash they had to themselves. Lending is the backbone of financial systems, and when it stopped, cracks started to appear in the system that had bound these banks together.

On August 9, 2007, a key financial indicator that showed how freely banks were lending to each other went haywire after a sustained period of absolutely no change. This meant that the financial sector was falling apart as banks tried to minimize risk and keep their cash for themselves. Another positive feedback loop was set up where banks started calling in loans from other banks they had lent to, which made the situations of those banks worse, requiring more loans to be called in until large parts of the banking system broke down, and some major institutions collapsed.

Sadly, these events were not confined to the big deals between financial institutions. Our money is also used in those deals, as were our mortgages. The contagion spread throughout the international financial system, affecting customers, pension funds, small businesses, large businesses, etc. With no more access to loans, many businesses collapsed, which worsened unemployment, which worsened mortgage defaults … you can see where that feedback loop is going.

The positive feedback loops only exhausted themselves when nearly everything caught in their loops had been destroyed or had become so devalued that they were practically worthless. The deals within deals within deals within deals meant there was no transparency in what was meant to be propping up those transactions. The hidden complexity of the system only began to manifest itself when the system collapsed.

4.5.1 Analysis

So much went wrong here that it is difficult to know where to start. The problems may have begun in the 1980s when computers were increasingly used in financial transactions. By the start of the 21st century, the entire global financial system was built around a computer-based infrastructure. While the advantages of this are self-evident, the problem was that the informational subsystem of the global financial system grew exponentially. Data were wrapped in data, which were wrapped in more data and then lost in a huge and ever-growing network of information. Metrics that would have been more prominent before the transformation of the financial system in the 1980s became lost in the maelstrom of noise from a planet full of computers analyzing, reanalyzing, predicting, etc. This increasing opacity of information also meant that less reputable practices could be more easily concealed within the data stream or explained away by blaming the increasing complexity of the informational system. Institutions might use collateral to support one deal and then use the same collateral to support another deal. What happens if both deals fail?

Increasingly, it seems that the structural issue was that the global financial system was all SYS1 and SYS2. Consider the days before widespread, rapid international financial transactions. Countries would have markets, currency, currency reserves, and central banks. Markets were regulated to some extent by the central bank, which then had access to the metrics they needed to set monetary policy and enact regulations. While it didn't always work perfectly, when problems arose, their effects were generally confined to that country. From this description, we can see that the financial system of a country is a viable system: the banks are the SYS1 entities, SYS2 and SYS3 enable the collection of economic data and the communication of policy, SYS4 sets policy, and SYS5 determines the overall goal of the system. SYS4 and SYS5 would be the central bank, the currency reserve, and the government working (hopefully) together.

Once financial institutions from different countries started trading with each other, the model changed. Through a process of autocatalysis, the global financial system started to build itself, though not completely. Each country's financial system became a SYS1 entity (remember, SYS1 is also a viable system because it also necessarily has SYS2-5). The interactions between them and the availability of huge amounts of data from abroad constituted a weak SYS2. However, there was nothing beyond SYS2 to keep the system in check. SYS1 entities are supposed to be able to act autonomously but with monitoring via SYS3 and then intervention from SYS4 down through SYS3 and SYS2 if disturbances start to occur. With the global financial crisis, there was no SYS3, 4, or 5 to intervene when the activities of the SYS1 entities began to generate massive but concealed problems which would ultimately almost crash the system. Central banks may be able to exert some measure of control over the financial systems in their own countries, but they can't control what is happening in other countries. For example, the banking system in the United States determines what can and can't be done with dollars in circulation between financial institutions in the United States. It cannot effectively determine how foreign banks use dollars that are outside of the US system.

From a systems perspective, we can see that the global financial system grew into a headless chicken. There was no Regulatory Body that could monitor and coordinate the activities between the subsystems that represented the financial institutions of different countries. There was no alert signaling network to make the operators in the system aware that something was going badly wrong, and it was only some people who managed to find that weak indication in all the noise that were able to make financial bets on the failure of the system which made them a lot of money in the end.

Because the global financial system didn't have the structure needed to ensure viability, even what was there was dysfunctional. SYS1 entities would trade with each other and would make trade-offs as well. Trade-offs can work in a system, but they need to be audited. With no audit, it was like these trade-offs were intersystemic trade-offs, which, from the Lake Peigneur case study, we know can be disastrous because they are unconstrained, i.e., no one has the responsibility or authority to decide whether these trade-offs are justifiable or whether they are going to shift the operating point toward a failure boundary.

SYS2 was flowing with information, but there was no SYS3 or SYS4-KNOWLEDGE to filter and crystallize the important information and make it available. Useful data got lost in the torrents of information generated by the computer-driven financial system every minute. Only those lucky few who managed to spot it got to benefit.

Once the system's disintegration started, the positive feedback loops that worsened the problem set in, and there was no Regulatory Body above that could intervene and stop them. Once the positive feedback loops had set in and the system's operating point was accelerating toward the failure boundary, a domain shift occurred from Complex to Chaotic. The response pattern needed to manage a system operating in the Chaotic domain is act-sense-respond, and there was no international organization that was empowered to act on the whole system. Central banks could act on the parts of the system they could control, but that was it.

Aside from the structural defects of the global financial system, the SYS1 entities and the "system" as a whole violated Ashby's Law of Requisite Variety. A system cannot exert control in its operating environment unless it can be in as many states as that environment can be in. Even without accounting for the fact that there was no Regulatory Body that could intervene in the global system, even at the level of the SYS1 entities in their respective countries, central banks only had very crude levers with which to control their systems. They could enact regulations (which could be ignored), control interest rates, manage the flow of currency into the markets, and use it to buy assets. These are relatively crude dials to try and maintain some measure of control over a hugely complex system. From the international perspective, there wasn't even anyone who could have used these dials had they even existed.

Perhaps the most troubling phenomenon with this global financial system is that even though it was essentially headless, the SYS1 entities and the financial institutions that they contained operated on the basis of competition. While competition and uncertainty are fundamental to financial markets, the lack of constraints on the degree of competition and the degree of uncertainty that central banks were willing to tolerate meant that this time bomb could be hidden within the financial system. The system would only become aware of it after it had exploded.

Chapter 4.6

Continental Airlines[1]

In the early 1990s, things for US carrier Continental Airlines were going from bad to worse. On top of record losses, the reputation of Continental was among the worst in the airline industry. The flying public actively avoided flying with them unless absolutely necessary, and morale among the staff was at an all-time low. The aircraft fleet was old, the interiors faded, and delays and technical problems were commonplace.

A management consultant brought into the company to try and effect a turnaround wrote about his experiences in a well-known article for the Harvard Business Review. It is well worth reading.

In the article, he comments on how dysfunctional Continental had become. Because of their poor performance, changes at the senior management level were frequent. Fortunately, Continental acquired a new Chief Operating Officer from Boeing just at the right time. Both men set about analyzing what structural problems Continental had and designing solutions to fix them.

Rather than going into all the things they did, I want to focus on one element. Morale among staff was low for many reasons, but one of these reasons was the extent of the procedures that they were supposed to follow. Over several years and several different senior management setups, the procedures for dealing with customers (and, possibly, dealing with other aspects of the operation) had become a huge, unwieldy mess. To try and cover every possible scenario, rules kept being added to the procedures manual to the extent that this voluminous tome was jokingly referred to as the "Thou Shalt Not" book. Staff knew it was not worth going above and beyond the call of duty to help a customer because they would probably fall foul of some procedure in the "Thou Shalt Not" book and open themselves up to criticism.

As part of the recovery plan, the new management team took the book of procedures to the parking lot and burnt it in a 55-gallon drum. Customer-facing staff were told that they were trusted to use their own initiative when dealing with customers with the aim of showing that Continental was a changed airline that now treated its clients with much more consideration and capability than in the past.

This, together with many other interventions, turned Continental into one of the most respected and profitable airlines in the industry, with the share price going from $2 to $50.

Designing and Managing Complex Systems. https://doi.org/10.1016/B978-0-323-91609-7.00018-4

4.6.1 Analysis

When something unexpected happens in a system, there is naturally a desire to try and understand it. When that event has been negative, there is naturally a desire to try and prevent it from happening again in the future. The problem starts when we try and manage the future by creating procedures.

Knowledge in a system can be explicit (written or recorded in some way) or tacit (stored in people's experiences and in their memories). In trying to constrain human performance within parameters that a management system considers acceptable, we must make sure that such knowledge is explicit: you will do it this way, or you won't do it that way. When something negative occurs, even just once, we may add in more procedures to constrain behavior even further. As this cycle continues, we might assume that we are making the system more reliable by more accurately and tightly controlling human performance, but, past a certain critical mass of explicit knowledge, the opposite thing starts to happen. Four phenomena lead to problems when systems become overproceduralized:

1. Learning capacity is exceeded – If procedures are too extensive, particularly if they are meant to be known rather than looked up in manuals as circumstances require, there will reach a point past which it is unreasonable to expect people to know them all. Even with a normal number of procedures in a system, there will always be a bell curve of performance with a few people knowing less than they should, a few knowing pretty much everything, and the majority knowing enough to get by. Finding out which are the procedures that everyone knows is a good way of seeing which procedures are most important. It may be that infrequently used procedures are more likely to be forgotten, and so the question for a manager is whether there is value in expecting people to know them all in the first place.
2. Contradictory information – With every procedure that is added, especially if this accumulation happens over time, there is an increasing chance that the procedures will begin to contradict each other. This introduces confusion into the system as employees now must find a way to resolve the incompatibility between procedures.
3. Outdated information – Changes happening elsewhere in the system may mean that an old procedure is no longer required. In some organizations, particularly safety-critical ones, it is relatively easy to add procedures into the system as they are viewed as legitimate ways of plugging perceived gaps in operational performance. However, removing procedures is far more difficult as it is harder to make a case to say that removing a procedure is better than adding a procedure. It can be especially difficult for a manager to be willing to remove a procedure that a previous manager had added for, presumably, good reason. One option is to consider the number

of procedures that we can reasonably expect employees to be able to handle. That number in itself then becomes part of the constraint on management and a way of keeping the operating point away from the boundary of unacceptable workload for employees. Suppose someone wants to add a procedure and that number has already been reached. In that case, they need to take another procedure out or find a different way of solving the particular problem that led to the procedure being written.

4. Time constraints – Generally speaking, if a procedure is being added to a sequential series of procedures that an employee is expected to perform, time needs to be added to the workflow as well. When we covered the concept of capability to perform functions, we identified time as being a resource. It is not uncommon for procedure after procedure to be added to a system without a corresponding increase in time available (or, indeed, any other resources needed) to complete the sequence of tasks. The efficiency/thoroughness trade-off decision is also made more complicated when procedures are added. It may be that more shortcuts are taken, and the gap between work-as-imagined and work-as-done starts to grow even further.

The addition of this type of component (procedures/software/laws) to a system is different from adding other components (humans or machines). Simply in the way that knowledge is stored within a system makes procedures a difficult component to manage. Sometimes they will be applied successfully; sometimes they will be applied incorrectly; and sometimes they will be forgotten altogether. In one of the models of sociotechnical systems we looked at in Part 2, the SHELL model, we saw that there were interfaces between how humans interacted with other humans, how humans interacted with machines, and how machines interacted with humans. In many systems, these interfaces are governed by procedures. An organization engaged in some type of production will likely have machines, and procedures often govern how humans interact with these machines. It may be in the way that humans program and maintain them, or it could be health and safety procedures that apply when working in proximity to them.

Instead of adding procedures to try and prop up a system that is having difficulties, we need to reexamine why these difficulties are occurring. Adding procedures should be a last resort when other, more permanent measures such as changing how a machine operates or increasing staffing to plug the gap in the operation aren't possible. If a procedure is added, it will require energy to train staff to follow it, and it is going to take energy to ensure that they continue to do so. If an audit discovers that there is already a gap between what you think people are doing and what they are actually doing, it will be more important to find out why this gap exists. Is it procedural drift, or is it an efficiency/thoroughness trade-off carried out to get the job done in the time available? If it does turn out that there is a gap, adding a new procedure may not to have the positive effect you think it will.

Finally, adding procedures is making a change. In complex systems, you cannot predict the consequences of changes once the system is operating. It may be that by adding a procedure to plug a perceived gap in an operation, another gap opens somewhere else, suggesting that the problem runs deeper than something minor that can be fixed by adding a rule. Expecting employees to take more time carrying out this procedure may mean that less time is spent on things that are ultimately more important.

Chapter 4.7

Three Mile Island[1–3]

In March 1979, an argument took place in an aircraft hangar in Pennsylvania. The argument was heated and was between two experts in some highly specialized field, each one arguing that their interpretation of events was the correct one. The stakes of this argument were about as high as you could imagine. If one of the experts was proved right, the nuclear reactor nearby was about to explode, spreading radioactive devastation across the United States in the same way as Chernobyl would spread radiation across Europe and Russia a few years later. To make the situation even more critical, the President of the United States was shortly to land at the airport before traveling to the plant to carry out an inspection. An explosion would then not only cause catastrophic radioactive fallout, but would also kill the President of the United States.

The problems had started a few days earlier at Unit 2 of the Three Mile Island nuclear power plant near Harrisburg, Pennsylvania. The problem seemed simple enough at first. Under normal circumstances, water passing through the reactor's core in the primary cooling system is heated to high temperatures. The pipework carrying this heated, radioactive water then comes into contact with the pipework of the secondary cooling system. Heat passes from the primary cooling system to the secondary cooling system, but no water is transferred between the two.

For this reason, the water in the primary cooling system is radioactive, but the water in the secondary cooling system is not. The water in the secondary cooling system must be extremely pure because when it turns to steam, it must drive carefully engineered turbine blades to generate electricity. Any impurities in the water would make this inefficient. A filter system keeps this water as pure as possible, although the filter in Unit 2 was known to be problematic and had failed several times recently. At 4 a.m. on March 28, 1979, a small amount of nonradioactive water from the secondary cooling system leaked out through the filter unit. Many of the instruments used to monitor the nuclear power plant rely on pneumatic pressure, and it seems that some of the water got into this air system and led to some instruments giving incorrect readings. One of the systems affected by this leak was the feedwater pumping system that moved the hot radioactive water in the primary cooling system into proximity of the cool, nonradioactive water in the secondary cooling system. The water contaminating the feedwater pumping system caused it to shut

Designing and Managing Complex Systems. https://doi.org/10.1016/B978-0-323-91609-7.00019-6

down, and with no heat being transferred, an automated safety device became active and shut down the turbine.

As well as circulating water in the primary cooling system so that heat transfer can occur into the secondary cooling system, the movement of water caused by the feedwater pumps was also what kept the temperature of the reactor's core under control. With these pumps shut down, emergency feedwater pumps were activated to keep the core temperature under control. For some reason the valves in both emergency feedwater systems had been closed for maintenance and had not been reopened. The emergency pumps were working, which was indicated in the control room, but the operators didn't realize that although the pumps were working, the valves were closed. This meant that feedwater couldn't get to the reactor to cool it down. There was an indication in the control room that the valves were closed, but one of these was obscured by a tag hanging from the panel above, and the default assumption of the operators was that these valves would be open as common knowledge had it that the valves are only closed during maintenance and testing.

With heat building in the core, the reactor automatically activated a system to stop further heat from being generated. Control rods dropped into the core to stop the nuclear chain reaction by absorbing the neutrons that normally sustain it. Even though the chain reaction had stopped, this huge stainless-steel reactor was still incredibly hot and required vast quantities of water to cool it to a safe level. Without this cooling, the high residual temperatures would continue to heat the remaining water in the core and could lead to a significant pressure build-up, enough to critically damage the reactor. To avoid this happening, an automatic pressure relief valve opened to allow some of the hot, radioactive water to leave the core and reduce the pressure. The pressure relief valve is only meant to be open for a short period to regulate the pressure. Unfortunately, the pressure relief valve failed to close after the pressure in the core had been reduced. Unbeknownst to the operators, the core essentially had a huge hole in it (the open relief valve). To add to the confusion in the control room, even though the valve was now stuck open, the system that would normally close it sent a signal to the control room saying that it had ordered the relief valve to close, and this was interpreted as confirmation that the relief valve was closed.

Problems continued to mount over the following days to the extent that a mass evacuation was ordered and churches in the local area offered general absolution to their congregations, something that is normally only offered when death is imminent. As all levels of local, state, and federal government tried to solve the problem, the situation deteriorated. The argument in the aircraft hangar was about whether a dangerous concentration of hydrogen gas was accumulating at the top of the reactor to such high pressures that it could detonate, blowing open the reactor core and spreading tons of nuclear material all over the United States. Fortunately for the population, it seemed that while hydrogen was accumulating, it was not at sufficient pressure to detonate spontaneously. By the time President Jimmy Carter had landed and was taken on a tour of the plant, an improvised system of pipes was feeding water to the core, thus cooling it, and the unexpected positions of the various valves had been discovered and corrected. Three Mile Island did not explode, but the reactor was critically damaged, and a substantial amount of the nuclear material had melted in the core, thus rendering it unusable. There had also been some release of nuclear material but not enough to pose a significant health risk.

4.7.1 Analysis

The nuclear accident at Three Mile Island was caused by informational failures. A predominantly technical system operating within the Complicated domain experienced informational failures that meant that the technical subsystems which should have been able to handle the small disturbance that started the chain of events, were incapable of responding appropriately. Once human operators began to get involved, a domain shift occurred. Had all the valves been in their correct positions and all the indicators accurately showed what position a valve was in rather than what position it was meant to be in, the automated safety systems would have activated when the feedwater pumps shut down, and the emergency feedwater pumps would have continued to cool the core. That would have been the end of the event. It was only because the valves for the emergency feedwater were shut, and this crucial information wasn't factored into the steps that followed that Three Mile Island found itself on the edge of catastrophe.

Although this case study is from a long time ago, it raises an important issue about the "socio" and the "technical" that go into sociotechnical systems. The interface between humans and machines has long been seen as a potential source of disturbances in systems. We might think we have given the correct instructions to a machine, but the machine does something we aren't expecting. The machine might "think" that it is telling us what we need to know in a way that we can understand, whereas it's actually masking the really important information behind a lot of less important information.

When it comes to problem-solving in complex sociotechnical systems such as nuclear power plants, the highly automated nature of such systems often makes it a lot more difficult for human operators to keep up with an evolving situation. From the first failure of the filter system and all the subsequent activations of automated safety devices, (for example, the emergency feedwater system and the pressure relief valve), only 13 seconds had passed. Not only was the plant in a much different configuration than it had been 13 seconds previously, but the problems with the various valves being in the incorrect position and this not being clearly signaled in the control room meant that the operators were completely in the dark regarding the true nature of the problem that they had to deal with. If the human and automated control mechanisms had been designed as a joint cognitive system, a system that recognizes that humans and computers operate at different speeds and so ensures that both components can keep up with the evolving situation, the human operators would have had a better chance of exerting control over the reactor.

In the 21st century, safety-critical machinery is often highly automated. A modern nuclear power plant would be expected to shut itself down and keep itself stable in an emergency. Although it is a highly interconnected system, it is still a technical system and should normally operate in the Complicated domain, i.e., a nuclear engineer should be able to visualize all the necessary connections in the system and predict their activity. Humans are involved but to a lesser extent than in facilities such as coal or gas-fired power stations. In those cases, failures tend to be self-limiting, whereas in a nuclear power plant, the function is to use the heat of the fission reaction to create electricity, but the plant must also function to prevent that reaction from generating too much heat even if electricity isn't being generated. This tight homeostatic control is most reliably achieved through automation.

More broadly, this case illustrates many of the technological, procedural, and human limitations that can be seen in just about every accident or incident in any industry. The human operators had to deal with a highly automated, complicated nuclear power plant that could radically change its configuration at a speed they couldn't keep up with. The indicators that they relied on to form an adequate mental model of the situation were unreliable, and many assumptions were made during the decision-making process that not only didn't fix the problem but made it worse. This accident also started in the early hours of the morning when humans are not best suited to complex problem-solving. Because of a limited number of phone lines running into the plant, it was almost impossible for people with vital information (such as the company that built the plant) to communicate this information to the people making the decisions. The subsequent investigation of this accident highlighted the vital importance of human performance in the face of complex, dynamic, novel situations involving sociotechnical systems. It also led to Charles Perrow developing his theory of "normal accidents", a theory which underpins modern thinking about the design and management of sociotechnical systems.

Chapter 4.8

Cybersyn and the trucking strike[1–3]

Some readers of a certain age may remember what a Telex machine looks like. For those of you who don't, it looked like a cross between an electric typewriter and a telephone. Telex machines were developed in the early 20th century to exploit established telephone lines to transmit text rather than speech. They grew in popularity during the mid-20th century until they were progressively replaced with fax machines and, later, email technology.

The benefits of the Telex machine were clear from the outset. Brief messages could be sent, which would then be printed out by the machine, thus providing a record. Later incarnations of the technology allowed a single message to be sent to multiple recipients, much like we would send a group email. Interestingly, some of the "text-speak" used today in text message communication originated with Telex machines.

In the 1970s, these machines would play a crucial role in the greatest attempt to apply the new science of cybernetics to a complex system. That system was the economy of Chile.

During the 1960s, Chile became an increasingly divided country. A growing and increasingly resentful political left-wing made up of the workers, professionals, and academics and an increasingly driven and defensive group on the political right who were the factory and business owners concerned with preserving private infrastructure and capital. The growing organization of the left along socialist lines later led to the election of President Salvador Allende in November 1970. A true socialist, he was committed to improving the quality of life for the workers and the poorer people in Chile. He resented that large swathes of those organizations that powered the Chilean economy were in private hands.

The rise of socialism in this South American country greatly concerned capitalist nations, especially the United States. American multinational companies had significant capital interests in Chile, and the rise of socialism put those investments at risk. To try and offset the progression of socialism in Chile, the US government under Richard Nixon directed the Central Intelligence Agency (CIA) to work with Chile's political right to disrupt the Allende government by whatever means necessary.

Designing and Managing Complex Systems. https://doi.org/10.1016/B978-0-323-91609-7.00020-2

Wealthy Chileans who owned capital investments in the country, and were mostly part of the political right, accepted money and training from the CIA to fund operations to undermine the socialist government.

While this activity was happening behind the scenes, Allende was looking for new ways to manage the Chilean economy more effectively. By renationalizing many industries, Allende was now faced with the problem of how to manage those industries.

One member of his government, Fernando Flores, was acquainted with the growing field of cybernetics and suggested that this could present a model of the sort of control system that they were looking for. As such, the Allende government contracted Stafford Beer, one of the early pioneers in cybernetics, to design a cybernetic management system to help manage the entire economy of Chile.

The geography of Chile itself presents its own problems. Being extremely long and thin with the Pacific on one side and variable terrain throughout, physical transportation within the country was difficult. Only one major highway linked the southern Patagonia region of Chile with the more populous north part. Constraints of geography meant that train lines were generally restricted to localities, and there was no train line running the country's length. Because of the lack of railroads, the trucking industry in Chile was essential for moving goods and supplies throughout the country.

The geography of Chile also presented difficulties in terms of communication. While communication networks in other countries might rely on a hub and spoke approach when it came to routing information, the only choice for Chile was a linear communication network running north to south.

Given the opportunity to build a cybernetic control system for the economy of an entire country, Stafford Beer worked with members of the Allende government, particularly Fernando Flores, to design a system that would be up to the task. This would become Project Cybersyn.

The plan was that an information network would span the country and would be able to feed back information about factory production to a central control room. Rather than being under bureaucratic control, these factories would be given more autonomy to control production as they saw fit. They would be able to use the communication network, planned to be called Cybernet, to communicate daily information about factory production and also convey requests for more resources when necessary. This information would then be fed into a computer model of the various types of industries in Chile, which would then be plugged into a broader representation of the entire economy of Chile. This would be based on Beer's Viable System Model, which was the basis of his theory of management cybernetics.

Once all the parts were connected, Project Cybersyn allowed people within the ministries responsible for industry and economy to quickly identify where there were product surpluses or shortages and would also allow them to react quickly to demands for more resources. Traditional models of economic management were based on the gradual collection and collation of statistics and their transmission to a central governing body. The slow pace at which this process would happen would mean that the people making the decisions were basing those decisions on information that was likely to be out of date.

As well as being able to see accurate, daily information about the state of Chile's various industries, the information presented through Project Cybersyn would also allow the managers of the economy to coordinate the means of transportation of materials between factories and, from there, to deliver goods to retail outlets or to locations where they could be prepared for export.

By 1972, the Cybersyn control room had been constructed in Santiago, the Chilean capital, and was waiting to be connected to the Cybernet communication network, which was still under construction. However, before the system became operational, political forces on the right backed by the CIA began to foment dissent among the remaining owners of private enterprises throughout Chile. In October 1972, in response to Allende's plan to introduce a nationalized trucking company, owners of private trucking companies initiated a strike.

Under normal circumstances, we consider strikes to be a political tactic used by organizations on the left, more so than by those right. This strike was unusual, and, arguably, without the backing of the CIA, it may never have happened. The people who initiated the strikes and shut down the factories were not the workers as one might expect in a normal strike. It was the business owners who initiated the strike. Initially, this happened in the transportation sector, but later, owners of businesses in other sectors joined in. This coordination of wealthier owners of businesses who occupied the political right effectively paralyzed the Chilean economy. If this had been allowed to continue, it could have been the end of the socialist government of Salvador Allende. Instead, something very unusual happened.

In response to this strange, inverted strike that business owners had organized, the political left mounted a coordinated effort to try and cancel out the effects of the strike. Even though the strike had been orchestrated by the owners of private industry who shut down their factories, the workers they employed went back to work anyway. They broke into factories, restarted production lines, and stole trucks so that they could continue to deliver goods and resources throughout the country. What should have been a crippling blow to the Chilean economy turned into a rallying cry for the workers and the socialist left to come together and try and keep the economy going in support of the Allende government.

Despite their best efforts, large parts of the Chilean economy continued to struggle, particularly with transportation. Only a fraction of the trucks that would normally be available to transport goods and materials throughout Chile were on the roads. To prevent this lack of transportation from affecting the economy at large, the Allende government needed to intervene and needed to intervene quickly.

Fernando Flores and Stafford Beer realized that the communication network they were building (Cybernet) could be brought into operation early to help Allende and his government organize the trucking resources that were available to them. Both had envisaged a more technologically advanced infrastructure for the Cybernet network, but because the trucking strike crisis happened so quickly, they realized that they had to use whatever technology was to hand to get the network operational. Fortunately, Chile had a large stock of Telex machines in reserve, and these were distributed to factories throughout the country. The Telex machines formed the backbone of the Cybernet network. As well as that, Fernando Flores set up a telephone network based on secret phone numbers that would allow socialist leaders in different regions of Chile to contact senior central government members in Santiago quickly and directly. This was what Stafford Beer would have called an algedonic signaling network; a rapid way of transmitting crucial information to the highest levels of power. In the revised VSM, this is referred to as the alert signaling network.

With the Telex machines now distributed and the Cybernet network up and running, Flores and Beer turned their attention to the trucking strike. They knew they now had a way of deploying trucks where they needed to go, but they also knew that they needed enough trucks to meet the transportation demands of the factories still in operation. In normal circumstances, not every truck is on the road constantly. It is normal to have a lot of redundancy in the system, so Flores and Beer could deploy these redundant resources to keep the country going.

The idea was simple. Factories would continue producing and would notify the central control room in Santiago via the Cybernet network when they needed more resources or when they had resources ready to distribute. This information could then be used to deploy the available trucks to keep the supply lines open. Because the situation was changing so rapidly, having information on what was actually happening being transmitted almost instantaneously to the people who were responsible for coordination was invaluable. Trucks could be routed efficiently to where they were needed, and efficiency could be improved by then seeing if those trucks could carry out any more transportation runs in that local area.

The use of the Telex machines also had another advantage. Because they printed out text messages on paper, there was an automatic record of the information received and the instructions that were being given. Suppose communication had just happened through ordinary phone lines. In that case, that information could likely have gotten lost in the system during the frenetic period when the central government was trying to alleviate the effects of this strike. With a paper record, it was possible to build up a more complete picture of what was really happening throughout the country.

The use of the Cybernet network arguably prevented the collapse of the Allende government in October of 1972. By being able to deploy resources quickly and accurately throughout the country, the government was able to keep the economy going despite the actions of groups on the political right. However, within Chile's population, the situation was not so clear. The workers and the socialist government had countered the strike organized by business owners, but such were the forces on each side, all that was achieved was a political stalemate. The work of the Cybersyn team and the use of the Cybernet network had kept the country going through the worst parts of the strike, but it would take something else to end the strike completely. To achieve this, Allende had to call on the military, who, in their turn, requested cabinet-level positions in the government for some of their officers.

From 1972 until 1973, the CIA continued their involvement in Chilean politics and backed the military junta, which would eventually overthrow the socialist government of Salvador Allende and install General Augusto Pinochet as leader. The brutal dictatorship, which marked Augusto Pinochet's time in office, led to the destruction of any work connected with Project Cybersyn. Socialist workers, as well as political and intellectual members of the left, were imprisoned and tortured. Public industry was privatized, and public investment plummeted. With the socialist movement effectively decapitated in Chile, foreign multinationals began to purchase land and businesses and would go on to invest more capital in Chile. Wages plummeted, and the benefits of these investments went to the business owners and the Pinochet regime.

4.8.1 Analysis

Project Cybersyn could, undoubtedly, have been the greatest achievement of cybernetics had it had an opportunity to be fully activated. Tragically, it was used as a pawn by the political right to try and convince people that the government was trying to control their lives using computers. However, Cybersyn and particularly the SYS2 of Cybersyn, Cybernet, did have an opportunity to show their value during the October Strike of 1972.

As you might expect for a system based on cybernetic principles, Cybersyn did contain most of the elements we see in the rVSM. Cybernet was the SYS2 that linked the various SYS1 entities that made up the Chilean economy with the necessary coordination and management systems (the Regulatory Body) in the Cybersyn control room and the governmental offices in Santiago.

When figures on the political right were trying to demonize Project Cybersyn and scare people into thinking that they would have their lives run by computers, the apparent structure of Cybersyn worked to their advantage. Simply by its arrangement, it does appear to be a hierarchical system. If you were not familiar with the concept, you might intuitively think that System 5 of Beer's original VSM is more important than System 1. It is, perhaps, unlucky that Beer chose to pick that particular way of representing the VSM diagrammatically. Clearly, he wanted to make it apparent that the model was based on the human nervous system which is, to an extent, hierarchical.

However, what cannot be represented on the diagram and is only revealed through reading and understanding the function of each system was that the viable systems that make up the different parts of System 1 (to use original VSM term) are the most important parts of the system. The rest of the system is there to ensure that the entities that make up System 1 operate as efficiently as possible. Despite the hysteria whipped up about computers taking over Chilean life, the VSM and, specifically, the intended autonomous dynamics of System 1 would have told the opposite story.

For clarity, I'm now going to talk in terms of the revised VSM rather than use the terminology of the original model.

SYS1 is based on autonomy. The best way for the whole system is to operate is if the subsystems that make up SYS1 are given broad autonomy in how they operate. Traditional hierarchical models would say that managers higher up dictate workers' activity lower down. Cybernetics says that the workers "lower down" have the best and most up-to-date understanding of what their part of the system is actually doing and what it actually needs. They also have the adaptive capacity to react quickly to changes in circumstances without recourse to management instruction from higher up.

If we fix this idea of broad autonomy as an essential feature of SYS1 entities, then we can get a clearer picture of what the purposes of SYS2, 3, 4, and 5 are; supportive metasystems for all SYS1 entities.

Telex machines, by virtue of the bandwidth available to them, require information to be concentrated. Operators within SYS1 would need to determine what information is relevant to SYS4 and SYS5 and then find some way of filtering out extraneous information and concentrating what was left into a transmissible format. If SYS2 had unlimited bandwidth, operators could potentially have sent all data they thought might be of interest to SYS4 and then relied on SYS3 and SYS4FOCUS to select the information they assess to be relevant to feed into SYS4MODEL. With the basic Telex system, the operator in SYS1 must make decisions that SYS3 and SYS4FOCUS would ordinarily make and also act as the transducer to turn that information into something that can be transmitted. In a fully operational rVSM, transduction occurs before filtering, and it is potentially less likely that important information will get filtered out before making it to SYS4MODEL.

Fernando Flores also did an excellent job setting up an alert signaling network. Given that Cybernet was being used almost exclusively to coordinate trucking movements, Flores knew that because of the widespread political unrest, it would be essential that local socialist leaders could raise the alarm if they saw the unrest escalating into something more severe.

All the parts were there and were working effectively. When it was put into operation early to deal with the strikes, the various parts hadn't been completely assembled and tested, but there was enough structure there for the government to mount an effective response to the strikes. Although this wasn't the purpose of Project Cybersyn, it demonstrated that the system worked. After

President Allende died, Stafford Beer returned to the United Kingdom. Many of the people who worked with him in Chile to set up Project Cybersyn were arrested, but Beer managed to get some of them released from prison using his network of international contacts.

The military destroyed the Cybersyn control room before there was a chance to see if the theory would work on an as yet untried scale; the whole economy. It remains to be seen if cybernetics can work on something so big.

Perhaps because of its association with Allende, Cybernetics began to be viewed as a manifestation of socialism. Beer himself produced a lot of work about the role of cybernetics in society and government. To take a dispassionate view, I think I would disagree with this position. Saying that the Viable System Model is an expression of socialism is akin to saying that your nervous system is an expression of socialism. A fascist government could organize itself by cybernetic principles and would probably find it was quite a successful way of managing things. Perhaps because it was a socialist government that was the first to see its potential, cybernetics got associated with socialism from the outset. It is also worth pointing out that since Project Cybersyn, management cybernetics has mainly been used within a capitalist framework of business development and enterprise architecture. I prefer to think about cybernetics as a generic tool for system engineering. Any political or philosophical interpretations are likely to come from the mindset of the commentator rather than from the system itself.

Chapter 4.9

Biological and informational viruses

When news of a mystery virus spreading through the Chinese city of Wuhan began to emerge in December 2019, there wasn't widespread panic throughout the rest of the world. Over the previous decade, pockets of viral illness had popped up in various locations worldwide, usually as the result of an animal disease crossing over into the human population. While some of these events may have been slightly more alarming than others, the pattern was usually the same; it would spread throughout a locality, cause some limited illness or occasional death, and then would burn itself out. Because of history, the world thought that the same pattern would repeat itself in Wuhan.

Why was Covid different? What was it that turned a localized outbreak of a subtype of the coronavirus family of viruses into a global pandemic? The spread and adaptation of the virus were also mirrored by another sort of epidemic that occurred right alongside—misinformation. As we look in more detail at these two stories, you'll be able to see where there are similarities in their dynamics.

A virus, like any other system, has a structure. That structure enables function, and the virus needs the capability to carry out that function. For the most part, the most important capability is access to a living host. Viruses are extremely simple organisms to the extent that there is an argument about whether they can actually be said to be alive. Most living organisms have the reproductive machinery to reproduce either asexually or through some exchange of genetic information with another member of the species.

Viruses reproduce in a very different way. Once a virus has invaded a cell, it hijacks the machinery that normally reads the genetic information in the cell's nucleus and turns that into proteins. Instead of making the proteins that the cell needs, this machinery now starts making more viruses. At some point, these viruses are released from the cell and can infect other cells. In other words, viruses need a host to be viable. Using a host to reproduce itself is the primary function of a virus, i.e., what it does.

For a virus, the success of their life cycle involves a bit of juggling. You need to get into another organism's cells to be really successful. You know that hijacking their cells isn't going to be too good for the organism, so you want to reproduce as much as possible without causing too much damage. If you cause

Designing and Managing Complex Systems. https://doi.org/10.1016/B978-0-323-91609-7.00021-4

too much damage too quickly, the organism might die before you've reproduced enough and gotten to the next step.

Once you've been replicating in the cells, you now need some way for these extra copies to go and infect another organism; otherwise, all this reproduction might be pointless if the organism eventually dies before you get to move on. Some viruses are clever and make sure that the damage they cause actually increases the chances that they will be transmitted, e.g., a respiratory virus incidentally causes coughing and sneezing, but these are both efficient ways of distributing viral copies to other organisms.

So far, you know that you don't want to be lethal, and you want to be able to spread. If you're smart, you might figure out a way of reproducing at low levels without causing too much damage so that the organism doesn't know it's infected. During this incubation period, they can be spreading copies of you very effectively.

Finally, there is your structure. Some viruses have a structure that means it is quite hard for them to get into cells and cause infection. Their structure will also affect how long they can survive outside of an organism and how widely the original host will distribute them. All these elements come together to give a virus its R-number.

The R-number was a statistic that cropped up in many news reports during the pandemic, and it is a useful aggregate parameter we can use to understand viral spread. Rather than trying to quantify all the biological parameters that contribute to a virus's transmissibility, the R-number simply looks at the number of people who are likely to be infected by a single person carrying the virus. This natural measure of transmissibility is sometimes called R-zero to distinguish it from the effective R-number of a virus that takes into account any preventative measures put in place, e.g., social distancing, lockdowns, etc. For our purposes, R-number is the natural transmissibility of a virus, i.e., in a general population that isn't taking any preventative action. If you think about it logically, a virus with an R-number of less than one will likely not cause an epidemic, given that each infected person isn't even guaranteed to infect one other person. If the R-number is greater than 1, this means that the virus will probably spread as each infected person is likely to infect more than one other person. For example, an R-number of 1.1 would mean that if there were 10 infected people, they would infect 11 people between them. If the R-number were 2, those 10 infected people would infect 20 people, and the virus would spread even quicker. The higher the R-number, the quicker the spread. The spread would also be exponential rather than linear. The case numbers for a virus that infects double the amount of people each time will not go upward regularly but will accelerate upward quickly as the infection rate doubles each time.

When it comes to the R-number, Covid distinguishes itself from other viruses that have caused outbreaks in the past. The first SARS coronavirus outbreak in 2002 was down to a virus that had an R-number of about 0.6 and led to a case fatality rate of about 15% of those infected. The MERS virus from 2012 had an R-number of 0.7 and a case fatality rate of 34%. While these fatality numbers are high, with an R-number below 1 and a high case fatality rate, these two viruses are limited in how effectively they can spread.[1]

The R-number for SARS-CoV-2, the virus that causes Covid-19, is approximately 3. That, combined with the low case fatality rate and a reasonably long incubation period, means that this virus is very well equipped to spread quickly through a population.

Viruses are also very adaptive. Given that viruses are reproducing so quickly and in such staggering numbers, the process of evolution happens at a dramatically increased rate. Evolution happens when a genetic mutation occurs during reproduction and, instead of being bad for the organism, it somehow gives it an advantage compared to other organisms that exist at the same time in the same environment. In animal evolution it might be that the mutation gives the animal's eyes improved sensitivity at night. This might make it more likely that they will be able to avoid predators and make it more likely that they will survive for longer and reproduce more successfully, thus passing on this advantageous genetic mutation to their offspring.

For mammals such as human beings, because our reproduction rate is so slow, these advantageous mutations are extremely rare, and this is why it took so long for us to evolve to have the structure we have now. Conversely, because viruses reproduce so quickly and on such a large scale, these mutations happen all the time. When a mutation happens that somehow tips the balance of those three transmissibility factors (ease of cell entry, incubation period, and ability to be widely distributed) in the virus's favor, that new variant of the virus now has a better chance of surviving. This is why we see new SARS-CoV-2 variants emerging over time.

Viral variants can be seen as complex systems, although not necessarily viable ones, given that there is a general lack of a Regulatory Body. However, in how successfully they spread, we can see that their structure is well suited to fulfill their function. Different variants can be viewed as competing systems much the same way as organizations can be competitors to other organizations. The variant that becomes dominant is the one that can deploy its adaptive capacity to gain an advantage. In this example, random mutations during reproduction are akin to adaptive capacity.

At the same time as Covid was spreading around the world, another epidemic was filtering out into the population. Throughout history, doubt and conjecture have accompanied disease whenever clearer, more scientific explanations have either been absent or have been misunderstood. It is only comparatively recently in human history that we have had the knowledge to describe infectious diseases in terms of infectious organisms transmitted among people. Before this, disease spreading in a population would be explained in terms of mythology, witchcraft, or religious punishment. As well as misinformation about the disease itself, vaccines have led to a huge increase in misinformation and have affected uptake rates internationally.

At this point, we have to recognize the distinction between misinformation and disinformation. Disinformation is the deliberate creation of false and misleading information. Misinformation is inaccurate information but may be transmitted by people who think it is correct. From these definitions, we can see that the creators of disinformation are distinct from the people who transmit it further down the chain of communication. It may also be that misinformation spreads because someone has unintentionally misinterpreted some fact but is not spreading the misinformation maliciously.

4.9.1 Analysis

In 2018, Prof. Heidi Larson, a professor of anthropology working at the London School of Hygiene and Tropical Medicine, wrote a prescient article in the renowned scientific journal Nature: "The biggest pandemic risk? Viral misinformation."[2] Given that it was published before the arrival of Covid, its conclusions were remarkably accurate. In the paper, Prof. Larson categorized types of misinformation according to how effectively it might influence other people's decisions:

1. Bad science – People with medical credentials trying to claim that a vaccine is dangerous without the scientific evidence to back up the claim. An example of this in the past was Dr. Andrew Wakefield claiming that the MMR vaccine (measles, mumps, and rubella) was linked to autism.
2. Financial opportunity – Stoking fear of vaccines while pushing consumers of this misinformation toward other products and services in order to make money. In this category, I would also include social media influencers who would use this or any other issue to increase their following.
3. Political opportunity – Making vaccines a wedge issue to divide a population and strengthen party political boundaries.

The Covid pandemic has borne out all of Prof. Larson's predictions: there are now groups of doctors who argue against vaccines; there are groups that actively spread misinformation to get people to buy their alternative therapies; vaccines have become a wedge issue, particularly in American politics.

The actual dynamics of viral misinformation share remarkable similarities with the dynamics of viral spread itself. In the same way as a virus has an R-number, misinformation also has an R-number based on factors affecting its transmissibility. These could be whether the misinformation is novel, credible, emotionally charged, and, more specifically, targets an external cause for the negative things that are happening or suggests some malign agency is behind everything.[3] This last point is important because it is a pernicious way viral misinformation can access our belief system by piggybacking in on the back of prejudice. Some misinformation about the virus itself suggested that foreign powers with evil plans created and released the virus to damage the economy of particular nations. Some misinformation about vaccines said that they were designed to control the population and that Bill Gates wanted to inject us with microchips.

Like the virus, successfully contagious misinformation has a way of accessing our thoughts to the extent that we may spread it further. The question is, how much further can one person spread misinformation?

For a virus to spread, infected hosts need to come into proximity with uninfected potential hosts. Someone with Covid isolating in a location by themselves is highly unlikely to infect anyone. Someone with Covid who is coughing and sneezing within a tightly packed, poorly ventilated bar is likely to infect lots of people. The same is true for misinformation.

This pandemic represented the perfect storm from a biological and an informational perspective. As we saw from the comparisons of the SARS-CoV-2 virus with other similar viruses, the combination of high R-number, low case fatality rate, and its incubation period made it an ideal virus to spread in our highly connected world. Additionally, compared to the state of the world and the nature of its communication networks during the Spanish Flu pandemic of 1918, the world has become a far more susceptible to viral misinformation because of social media

A prolific social media user whose updates are deemed "relevant" by the algorithms behind the platform can quickly become a super spreader of misinformation. As their social media followers pick up this misinformation, they can spread it to their followers. As with viruses, the spread is determined by reach; the more people you are in contact with, the more quickly and more widely you can spread the virus/misinformation. In the early days of the pandemic, this transmission of misinformation was largely uncontrolled. It was only later that social media platforms began to edit content to try and limit the spread of misinformation.

The algorithms underlying various platforms have also been implicated in creating echo chambers where people prone to believing misinformation are preferentially exposed to posts from other users sharing the same views. This has the effect of strengthening the perceived accuracy of the misinformation while limiting the opportunity to be exposed to counterarguments. These echo chambers become "epistemic bubbles" where knowledge that does not concur with that inside the bubble cannot gain access. Research has shown that Facebook and Twitter's algorithms are highly likely to create these epistemic bubbles.[4] For users, unless their real-world exposure to information can somehow counter this effect, the only knowledge they may have access to is misinformation. As with a viral contagion, as the misinformation spreads, these epistemic bubbles start appearing and increasing in size, much like individual viral outbreaks occur when infected people travel between different locations.

From immunological studies, it seems that individual susceptibility to various viral infections is not uniform. Some people are more likely to be infected with viruses than other people. On top of that, people's reactions to infections vary widely also. Some people might suffer a very mild illness, while some will die. Our immunological reactions to potentially threatening organisms are highly variable.

In the same way, people's reactions when exposed to misinformation vary as well. Some people are naturally skeptical of information that they receive through certain channels, e.g., online. Much like human beings' innate immune system (the part of our immune system that attacks any invading organism as quickly as possible), the person can immediately discount information that seems suspect.

For other people, they might briefly consider the content of the misinformation to assess whether they think it is accurate. For example, people with some sort of scientific or engineering background would quickly be able to tell that the idea of microchips in vaccines is highly unlikely. Given that they have got as far as considering it, we can say that it has accessed their consciousness but is almost immediately fought off by previous knowledge. This is much like the second step of an immune reaction where the body's acquired immune system uses biological information that it has learned during previous infections (such as the best antibodies to use to the fight the infection) to mount a more targeted immune response.

The problem lies in people who have no innate and no acquired immune system with which to mount a rational response to potential misinformation. Those existing in epistemic bubbles will be particularly prone to this because they are not exposed to any contrary arguments.

Even if they were, the phenomenon of asymmetric belief updating (which is similar to the confirmation bias that can happen in human decision-making) makes it less likely that new evidence, which is contrary to what they believe already, will actually update a person's belief. Evidence that seems to confirm a person's belief will likely strengthen that belief to a much greater degree than evidence to the contrary will weaken it.[3]

The informational systems that crisscross our world to an ever-increasing extent are a fertile breeding ground for these misinformational viruses. We also know that some people are more susceptible than others and are potentially more likely to suffer harm by believing this misinformation. Given that we have developed some approaches for managing the dynamics of biological viruses to limit their spread and mitigate any damage that they cause, are there lessons that we can apply to combat misinformational viruses?

The most effective measures that helped combat the spread and limit the damage of Covid have been vaccinations, social distancing, localized lockdowns, and the isolation of infected individuals. Can a consilient approach to managing misinformation strengthen the informational system that we use to build our beliefs about the world?

If we consider vaccination, it might be possible to inculcate more skepticism on the part of social media users through education. In a "post-factual" world, it is increasingly difficult to know for sure whether the information you are receiving is accurate, and so the individual must work harder to make that assessment each time they see a new piece of information. A lot of research has been carried out along these lines, and certain approaches seem to demonstrate that this sort of "vaccination" is possible. In a fascinating research study, Prof. Sander van der Linden, a social psychologist working at Cambridge University, developed a game with his team that encourages participants to create example of disinformation as a way of "vaccinating" them against disinformation/misinformation that they may encounter online. In the game (www.goviralgame.com), players are introduced to the strategies that creators of disinformation exploit to increase the likelihood of it spreading. These strategies include using emotionally charged language, referring to fake "experts," and suggesting that there is a conspiracy going on. Evidence shows that by exposing people to these strategies, they become more adept at spotting disinformation/misinformation in the future.[5]

Educating people about the nature of these social media platforms' algorithms could be helpful, but not as helpful as changing the algorithms themselves to limit the creation of these epistemic bubbles. The motivations for social media platforms to present users with evidence that constantly confirms their beliefs are complex. It is likely that the commercial drive is to keep users online for as long as possible to maximize the time that they can be exposed to advertisements. By configuring a newsfeed to present only information that reinforces the beliefs that users already hold (which could be ascertained by algorithmic programs that look at what comments and news stories that person has liked or reposted), they are likely to stay online and interact further with the platform and the associated advertising.

If this is true, the commercial nature of the organizational systems behind these social media platforms has created an informational system that is designed to form epistemic bubbles. This, combined with asymmetric belief updating, is a powerful driver of both the creation and the spread of misinformation and a powerful force that keeps people within epistemic bubbles.

There are undoubtedly freedom of speech implications when it comes to any suggestion about more actively managing the informational systems around us. By restricting someone's access to a particular platform as a way of "isolating" someone who is spreading misinformation, are they being censored? Again, this is a complex issue, but in the same way as the current informational system has been designed to encourage epistemic bubbles, it could potentially be adapted to prevent them. Is it right that a single person, by virtue of fame, or number of followers, can inject information directly into the newsfeeds of millions of other people in a completely unedited manner? Traditional news sources are regulated, to varying degrees, by laws and journalistic ethics. These types of mass media are also limited in how often and in how individualized a format their information is communicated to consumers. The depth of cognitive processing of something seen on a news broadcast and something seen in a social media news feed is likely to be different. So too will be the degree of individualization of the information flow, given that social media algorithms are normally based around "relevance" to the user in a way that regular news broadcasts can't be.

During the Covid pandemic, we've seen how susceptible the informational system that we use to create our beliefs is to the effects of disinformation and misinformation. Unfortunately, the high coupling and interrelatedness of this system that exists between media consumers, particularly social media consumers, means that misinformational viruses can spread more quickly than biological viruses. However, it may be that by adapting the techniques that humans developed to limit the spread of the SARS-CoV-2 virus, we can improve the integrity of our informational systems to give people more faith that what they are seeing is likely to be the truth.

Chapter 4.10

Netflix[1]

In researching this case study, I came across an article from 2014 entitled "The rise and fall of Netflix: what happened and where will it go from here?"[2]

When this article was written (probably around 2012 based on the financial data used), Netflix was struggling. Some management decisions and some unsuccessful changes to their operating system left them in a financially precarious position with a dropping share price. Hopefully, the authors saw the potential and bought a few shares as, from 2012 to 2021, they went up by 6000%, although the price then dropped sharply in 2022.

Launched in 1998 in California, Netflix started as a DVD rental company that used postal delivery instead of the normal rental outlet locations used by competitors such as Blockbuster. Initially, they worked on a pay-per-rental model before switching to a subscription service a year later.

In 2002, the company went public and used the money to expand its operation, reaching 1 million subscribers by 2002. In 2007, Netflix decided they would join the growing streaming market and started producing their own content two years later. After several more years of growth, the company expanded into South America in 2011, a year that proved extremely turbulent for the company.

In 2011, the CEO announced that Netflix would be splitting its DVD rental and streaming services into separate companies. Clients who wanted to continue to have access to both services saw their monthly subscriptions go up by 60%. The company share price dropped by 75% as 800,000 subscribers quit. On top of that, Netflix suffered system failures that paralyzed the streaming services.

Despite these issues, in the following year, Netflix consolidated its customer base and its technological infrastructure. 2015 marked the launching point of a meteoric rise in share price. International expansion brought in more subscribers, and by 2021, 200 million people worldwide had a Netflix subscription, with the share price peaking at approximately $700. Netflix's original content is often critically acclaimed, and its film, "The Trial of the Chicago 7," won seven Academy Awards, including Best Picture. More recently, the Netflix share price has dropped dramatically and it remains to be seen if they can recover as successfully as they have done in the past.

Designing and Managing Complex Systems. https://doi.org/10.1016/B978-0-323-91609-7.00022-6

4.10.1 Analysis

The Netflix story is one of evolution by natural selection. It is not the size of the company today that distinguishes it from its early beginnings but rather what it is doing now compared to in its early days. By evolving in tandem with its operating environment (especially with regard to the huge improvements in broadband Internet infrastructure), Netflix has been able to shed product lines that were no longer needed (DVD rental) and redirect those resources toward its now core business model: the provision of online streaming services.

As well as taking advantage of the changing technological landscape, Netflix has also leveraged its position as a leading entertainment platform by marketing itself aggressively through social media platforms. By providing previews and samples through social media, Netflix has built up a considerable following by focusing on a target audience of current and potential users. Again, this demonstrates how the organization has quickly adapted to changes in its operating environment, given that a large proportion of social media users will likely be open to a web-based entertainment platform. In this way, Netflix's growth was closely associated with the growth of other web-based service platforms such as Facebook, Google, and Amazon.

Realizing that content providers were not always reliable in allowing third parties to access their material, Netflix started creating its own. A new SYS1 entity was created whose output formed the input to another SYS1 entity: the streaming infrastructure. In 2020, Netflix original content accounted for approximately 35% of the content available on the platform.[3]

The in-house data analytic processes are proprietary, but they would surely give an insight into the function of SYS2 and SYS3 in coordinating the two SYS1 entities that deal with production and streaming respectively. It is likely that the nature of the content produced is carefully planned to account for the viewing wishes of the company's subscribers, all the data for which would be readily available. In traditional content-producing organizations such as movie studios, ideas are formulated and pitched to studio executives who decide whether they think they will succeed. Netflix's former head of original programming, Cindy Holland, has said that Netflix Originals takes a far more algorithmic approach to their decision-making given the quantity of data they have collected over the years that can give an insight into what is popular, and how that popularity varies across demographic groups. This is another example of how algorithmic decision-making often proves superior to human decision-making with its many biases and inaccuracies.[4]

These same algorithms allow for the personalization of content presented to users on their homepage and the personalization of marketing via the various platforms used for that purpose. Like other successful organizations, Netflix is very successful at leveraging all the available data to refine its operating processes and satisfy customer needs. As we will see in another case study involving deep learning computer algorithms, this relationship between

SYS4LEARN and SYS4KNOWLEDGE is likely to become far more prevalent in the future across a broad range of different organization types.

From that article published in 2014, it would be interesting to see what the next decade holds for Netflix. The effects of the Covid lockdown may have driven subscribers up, but will they continue to use the platform as regularly? A drop in subscribers in 2022 would suggest not. With more reliable and faster broadband infrastructure being established throughout the world, not to mention 5G connectivity that can support streaming services, Netflix's growth may restart again. The growth of deep learning algorithms is theorized to lead to scriptwriting programs that are calibrated to what will be popular with certain audiences. Already, analytic software companies such as Scriptbook can assign parameters to scripts that can guide producers in a way that is quantitative and reproducible rather than qualitative and subjective.[5]

It may be that the future of entertainment systems will be based on the algorithms they develop so that they are better able to give the user what they want.

Chapter 4.11

Fukushima[1–5]

Masao Yoshida was about to do something that wasn't supposed to happen in Japan. He was going to disobey his boss. Later, when his boss asked him what he had done, he would lie about it. This is very unusual in Japanese corporate culture, where hierarchical structures are considered to be important, and subordinates are expected to show superiors appropriate respect and deference.

Yoshida had decided that his disobedience was necessary because he knew more than his bosses did and thought his boss's motivations might have been different from his. He was trying to stop a nuclear meltdown.

On March 11, 2011, an earthquake registering 9 on the Richter scale struck off the Japanese coast. This was the largest earthquake Japan had experienced for 1000 years and was so strong that it actually displaced the coastline 8 feet eastward.

So large was the displacement of the seabed that a huge tsunami formed, which then headed for the coast. As the seabed shallowed, the height of the tsunami grew. By the time it arrived at the coastal site of the Fukushima nuclear plant, it was already high enough, approximately 40 feet, to breach the sea wall defenses and flood the plant. The tsunami was so powerful that it traveled 6 miles inland.

When the earthquake hit the coast, the emergency systems at the plant automatically shut down the reactors by inserting the control rods, which stop the fissioning process in the uranium. However, the reactors still retain a lot of residual heat even though the fission reaction has stopped. For this reason, they need to be continuously cooled until all the residual heat has dissipated. Reactors 1, 2, and 3 had cooling systems that would remove this residual heat, but the flooding of the plant meant the loss of electrical power. The emergency diesel generators were located underground and were also rendered inoperable by the flooding. The cooling systems of these three reactors gradually began to fail, and the residual heat began to cause the cores of these reactors to go into meltdown. Fortunately, the design of these reactors accounted for this possibility, and it appears that the thick concrete layer surrounding the central core was sufficient to contain the lava-like radioactive material and stop it from burning through into the ground and contaminating a wide area.

Unfortunately, part of the cladding on the uranium rods began to break down, which created a large amount of the flammable gas hydrogen. This leaked into the outer reactor building and reacted explosively with the oxygen in the air. One by one, the reactor buildings of reactors 1, 2, and 3 exploded, although the central

Designing and Managing Complex Systems. https://doi.org/10.1016/B978-0-323-91609-7.00023-8

reactor chambers weren't breached. Even though the cores weren't breached, a large amount of radioactive material was ejected into the atmosphere.

As if this wasn't enough, in addition to the uranium in the cores, uranium was also stored in pools of water in each reactor building. These pools would routinely be used to store spent fuel rods that had been withdrawn from the reactors themselves. Like the uranium in the cores, these control rods also had residual heat, and with the loss of the cooling systems, the water in the pools was beginning to boil.

The amount of fuel in the storage pool in reactor 4 was equivalent to all the fuel in the cores of reactors 1, 2, and 3, with the added disadvantage that it wasn't surrounded by the steel and concrete shell like the cores were.

With only torches and lamps, a small group of people had to improvise ways of keeping the storage pools and reactor cores supplied with water to keep them cool. Masao Yoshida, the plant supervisor, decided to use seawater to cool the reactors and the storage pools. He and his team used every possible method to get this seawater into the cores and storage pools, even using fire trucks to carry out the pumping.

His bosses, fearing the release of contaminated water into the sea, ordered him to stop until they could get approval from the Prime Minister of Japan. Knowing that the release of contaminated water posed a lower threat than the meltdown of the rods in the storage pool, Yoshida decided to continue pumping the seawater into the cooling system. Knowing that his bosses would replace him if they found out that he had disobeyed their order, he lied to them and told them that the seawater pumping had stopped.

This decision likely helped limit the release of radioactive material into the environment, although it would take many more days to get all the reactors cooled down to the extent that no further damage was going to occur. His bosses later reprimanded Yoshida for disobeying their orders, but expert analysis after the event suggested that his decision prevented the situation from being much worse. The cleanup operation is likely to continue for decades to come.

The Onagawa nuclear power plant had started operating later than Fukushima. The first of its three reactors came online in 1984. This plant was located significantly closer to the epicenter of the earthquake and was hit by a tsunami as high as the one that had flooded Fukushima. The difference was that during the construction of the plant, they had accounted for the historic possibility that a 40-foot tsunami could hit the eastern coast of Japan and so designed their seawater defenses to protect against nearly a 50-foot wave. While the plant experienced some damage, there was no release of radiation. Such was the level of defense afforded by the plant's designers, local people actually came to the plant and sheltered in the gymnasium. The plant was shut down for a long period, but permission has been granted to restart operations. In response to the height of the 2011 tsunami, they have increased the height of their protective sea wall even further.

4.11.1 Analysis

In some ways, we can take the perspective that there was a success story within the Fukushima disaster in so far as a much worse event was avoided, thanks to the actions of Masao Yoshida and his team. Unfortunately, their efforts happened against a background of regulatory failure and design inadequacy. Reports published by various agencies after the disaster highlighted deficiencies in design that contributed to the flooding of the plant. The nuclear power plant at Fukushima had been in operation for about 40 years. The sea walls surrounding it were 19 feet high. Some reports noted that there had been evidence that Japan's eastern coast had experienced a higher tsunami than this in approximately the 9th century, but the research that highlighted this wasn't published until 2001, well after the plant was up and running. When evidence came to light that a higher tsunami had hit the coast, the management at the plant thought it unnecessary to augment the flood defenses based on a once-in-a-millennium event.

The decision to locate the emergency diesel generators in a place where they would be highly prone to damage in the event of flooding was also seen as a failure of regulatory oversight as it was the loss of these generators that crippled the plant's ability to keep cooling water circulating through the building. However, moving the generators once they had been installed would have been very expensive.

Despite these failures and the ultimate destruction of the plant, the actions of Masao Yoshida demonstrated that even in a highly hierarchical system, autonomy could be asserted.

Given the extent of the disturbance that this system encountered, the desire to maintain hierarchical control was unproductive. With the loss of electrical power at the plant, the idea that plant workers could convey the complexity of the rapidly unfolding situation to their superiors at headquarters and allow them to control the repair efforts was unrealistic. SYS2 and SYS3 were seriously impaired, and rather than increase the autonomy of the SYS1 agents who were in the plant, SYS4 tried to maintain control. Because of the serious nature of the consequences, they also felt they had to get permission from governmental departments despite these being separated from the main operators in this situation by both geography and experience in nuclear technology.

There is an argument that the company that operated the power plant was keen to defer decisions that might have resulted in the release of nuclear material to the government to limit their liability in the future. This may or may not be the case. What is clear is that decision-making not only remained hierarchical, but an extra level of hierarchy was added (the government) to an already dysfunctional control system.

Chapter 4.12

The Mumbai Dabbawalas[1,2]

Every working day in Mumbai, a team of 5000 people is responsible for delivering 130,000 meals within the space of an hour. Despite relying on handcarts, bikes, and the occasionally unreliable train service in Mumbai, evidence suggests that 99.9999% of deliveries occur accurately and on time. Such is their success that the Mumbai Dabbawalas are recognized by high-reliability development organizations such as Six Sigma and the International Organization of Standardization.

Starting at 9.30 in the morning, dabbawalas, the people who collect the homecooked meals, visit all the homes of the clients in their district. The meals are already packed into uniform lunch tins known as dabbas. Whoever has prepared the meal needs to be ready to hand the tin over as soon as the dabbawala arrives at the door. If a client is late three times, they no longer can use the dabbawala service.

The dabbas from the homes in the district are packed tightly onto handcarts or fixed to pedal bicycles and are conveyed to the nearest train station. Here, they are sorted according to the destination station using an ingenious coding system painted on the lid of each dabba. Once sorted, they are packed into the cargo carriage of trains heading toward the city. Different trays are removed at different stations by waiting teams of dabbawalas.

At the destination station, they are sorted again according to the district and building that the dabba is bound for. Once again, they are loaded onto handcarts or bicycles and are delivered to their destination.

Such is the efficiency of the system that it is almost unheard of for errors to occur in delivery. Dabbawalas are not allowed to eat their lunches until all dabbas have been delivered to their clients working in various locations around Mumbai.

Homecooked meals are popular in Mumbai as it enables people to avoid the time and expense of eating lunch away from their workplace each day. The severe overcrowding on Mumbai's transportation network means it is easier to travel to and from work without carrying the dabba, and hence, the dabbawala system was born.

After lunch, the dabbas are collected from their clients at their workplaces and returned to their homes by reversing the system. The next day, the process starts all over again.

Designing and Managing Complex Systems. https://doi.org/10.1016/B978-0-323-91609-7.00024-X

4.12.1 Analysis

The dabbawala system of Mumbai has been studied by various business education institutions to try and work out the reason for its high reliability and low price point. The organization has a generally flat management structure with teams of about 25 dabbawalas working together in each particular region. These teams have a large degree of autonomy when it comes to finding customers, dealing with sickness or absence, or finding alternative means of transport in the event of delays or breakdowns on the rail service. Surprisingly, there is almost no technology involved in this setup.

The autonomy they have to solve problems as they encounter them is built on the fact that they have the necessary experience of the city, particularly the local areas that they work in, and they know the various choke points of its transportation infrastructure and so are able to quickly reroute dabbas if the normal transportation route becomes blocked.

From the revised Viable System Model, we know that SYS5 comprises the organization's overall function/goal and overall ethos. In the dabbawalas case, their organization has one function: the efficient and profitable transfer of dabbas from clients' homes to their workplaces and back again. Over the years, suggestions have been made that they should expand this function and start offering to cook meals themselves and so become caterers. Although the managers in the organization consider ideas like this when they arise, they are reluctant to change a winning business model. Instead, expansion comes through recruiting more clients and expanding geographically rather than moving into different business sectors.

More interestingly, the organizational ethos gives a real insight into the system's success and why the apparent lack of diversity in the workforce seems to work in its favor. The dabbawala system is based on four pillars: organization, management, process, and culture. From their remarkable efficiency rates, it is evident that the first three have been well implemented. It is the last pillar, culture that gives more perspective on why they are so successful.

The dabbawala organization first came into being around 130 years ago. Since then, expansion has led to an almost religious feeling among its staff in the importance of their mission. Delivery of food is likened to delivery of medicine; providing people with homecooked meals improves their day and improves their lives.

While most organizations, and even this book, promote diversity as an important factor in creating flexible, responsive organizations, the dabbawala setup does a good job of refuting this. Despite the diversity in Mumbai as a city, the dabbawalas generally belong to the same socioeconomic class, have a similar level of education, and share the same faith. As they may work in small local teams throughout their lives, these people from similar backgrounds have one more thing that cements them together as a community: they are all dabbawalas. Teams that also act as supportive communities to the members will be more likely to show reciprocity than teams of people with no common thread binding them and who just happen to end up working together. This is similar to the relationships that form within the subaks of Balinese rice farmers. Management textbooks will highlight that diversity allows for novel solutions to novel problems, but in the case of this system, this may be less important than accomplishing the simple stated goal of the organization with a high degree of reliability. Autonomy still exists to solve problems around transportation, but the generation of solutions is more likely to flow from deep familiarity with the urban area in question rather than novel, out-of-the-box problem-solving.

Again, in the world of complex systems, we come across the issue of balance. How do we balance the desire for a flexible, agile workforce who are constantly adapting to problems and spotting opportunities with that of having a homogenous workforce who have an easier time bonding as a team and supporting each other when necessary? For this particular system, the latter seems to work better. It may be because the organization's function is relatively simple despite the complexities of the environment. However, the complexities of the environment that are most likely to disturb this system are ones of transportation, and with their innate knowledge of Mumbai, this group of people has all the adaptive capacity that they need.

Chapter 4.13

Flash Crash[1–4]

When the financial markets opened on May 6, 2010, it looked like it would be a moderately turbulent but otherwise unremarkable day. American markets weren't doing great because this was the period when it looked like Greece might default on some of its debt repayments, and there was concern that this might destabilize other parts of the European markets.

At 2.42 p.m., something unexpected happened. The Dow Jones index dropped by 9% within minutes. No financially significant news prompted this huge drop in the market. Such drops are associated with sudden, market-altering events such as the collapse of a government or some natural disaster. In this case, nothing had happened in the outside world that could explain why so many shares were being sold, thus driving the prices and the market downwards. Within 20 minutes, the market had rebounded to where it was before. Again, nothing newsworthy in the outside world was responsible for this. The whole crash was over very quickly, but for those 20 minutes about $1 trillion disappeared from the markets.

Something very worrying had happened. What was more worrying was what would have happened if the market hadn't corrected itself? With nothing seeming to drive the huge sell-off of so many shares and nothing (except the lower price) seeming to explain why the markets jumped up again so quickly, it was conceivable that the same phenomenon might keep driving the market down without giving it a chance to recover.

Several agencies launched investigations into what had happened. The rapid and coordinated nature of the sales suggested that it may have been down to a computer-driven trading program used by High-Frequency Traders (HFTs).

HFTs use sophisticated computer systems to make and then adjust multiple orders for buying and selling stocks based on market movements. Unlike traditional investors who might focus on a company's fundamental commercial potential, algorithmic traders use data analytics to examine second-by-second changes in the market to identify opportunities to make a short-term gain. These types of trading programs work extremely quickly to try and take advantage of small moves in the market to make a profit. When volatility in the market increases too much and these initial moves can't be predicted, these programs will generally sell their positions and wait for volatility to subside. HFTs also use other, less scrupulous tactics to make a profit.

Designing and Managing Complex Systems. https://doi.org/10.1016/B978-0-323-91609-7.00025-1

Spoofing is the act of placing a large order into the market with no desire to have it filled (i.e., it might be an order to buy or sell but the person who places the order has no intention of buying or selling anything). Because stock prices move based on demand, the number of orders to buy or sell moves the stock price. Placing a large order can be enough to move the stock price in one direction. The order is then quickly canceled, but a profit can be made because the operator now knows in what direction the market is going to move.

Several other techniques can be used to move markets one way or the other, but this type of market manipulation is illegal, and the people who monitor the markets are supposed to crack down on it. However, because so much of it happens so quickly, it was relatively rare for anyone to be prosecuted for it.

Another feature of HFTs is that they are, by necessity, algorithmic. Because trades are executed so frequently, no human being could manually track or enter those orders at such high speeds. Computer programs are given broad scope to trade autonomously, and inevitably, computers then end up trading with each other, sometimes at incredibly high speed.

The cause of the Flash Crash remains controversial. At one point, it looked like a single trader operating out his bedroom in his parent's house in West London might have been responsible. This trader, after years of trying to get regulators to reign in the practices of the HFTs, decided to insert a bit of disruption of his own into the markets to try and teach them a lesson. Using a combination of techniques, most of which were against market regulations, he created an environment that made it more difficult for HFTs to turn a profit. At the same time, he also made remarkable profits for himself.

The market conditions of May 6th made it ideal for quickly turning multiple small profits. With markets swinging upward and downward, with the right timing, a savvy trader could get in for a few seconds, make a trade, let the market move, and then cash in their position. The trader in his bedroom in London had been doing this all morning and had managed to accumulate approximately $860,000 during the day. However, he had logged off before the Flash Crash happened.

In the postevent analysis, investigators suggested that his activities had potentially worsened the volatility that day, but they couldn't say that his activities actually caused the crash.

It transpired that there had been what we could comfortably call "a disturbance" just before the crash. A large mutual fund, a financial organization controlling a lot of money in pension funds and other investment vehicles, decided to sell a large block of futures contracts. Think of this like selling shares. The principle is almost the same. The amount that they wanted to sell had a value of approximately $4.1 billion. Rather than dumping all of this on the market at once and potentially reducing the price, traders normally execute a large order like this over time. A similarly sized trade earlier in the year had taken five hours of gradual selling before it was completed.

On May 6th, on a background of intense volatility, this sell order was placed on the market with no time or price restrictions that would limit how quickly the entire trade was completed. As such, over approximately 20 minutes, $4.1 billion worth of these contracts flooded the market and drove the price down.

HFT algorithms quickly bought a lot of these contracts at the depressed price, but as the market dropped, they switched to selling them. This effect oscillated between the various HFTs leading to some truly unusual, computer-driven market behavior. Huge numbers of contracts were bought and sold between computers. It is important to know that these transactions weren't being done with a view to keeping the contracts but rather to buy or sell them again as soon as the price moved slightly. The rate of transactions spiraled upward as these contracts were handed around in a fashion described in the postevent report as a "hot potato." Soon, the majority of these contracts were locked up in ever-increasingly complex trades, and the liquidity, (the number of contracts available for everyone else to buy and sell), dropped.

Such was the extent of the mayhem on this particular market that it would have been impossible for it not to spread. Rapidly, other markets began operating erratically. Trades began to happen at outrageously bizarre prices. At one time, shares normally valued at $50 changed hands for 0.0001 cents, and other mid-range shares were transacted at $100,000.

Markets have an inbuilt circuit breaker that stops trading if it looks like something truly irrational is occurring. These circuit breakers activated, and that may have been enough to break the cycle of irrational, high-frequency trading between the computer programs. The markets gradually recovered, and after about 30 min, the crash was over. Anyone looking at the graph showing market prices for the day could not help but spot a large V-shaped gouge that happened in the early afternoon.

4.13.1 Analysis

The exact causes of the Flash Crash are still unclear. Official reports suggested that the large sell trade from the mutual fund triggered the event, but this conclusion wasn't universally accepted. Critics pointed out that even though it was a large trade, it was still relatively small compared to the number of trades that might ordinarily be expected to happen on a market of that size in any one day. It may have been that multiple things were going wrong on that day, and they just happened to all line up within a narrow time frame and cause the crash.

With the huge increase in the number of computers undertaking market transactions, the market has lost the traditional cognitive limits that might have halted a crash like this much earlier on if it had been down to old-fashioned voice trades happening in chaotic trading pits between human traders. Only so many trades can be shouted at any one time, and so it would have taken longer for the downward trend to appear, and traders would have had time to consider what was happening before trading further. The "hot potato" phenomenon where large numbers of contracts were being rapidly circulated between computers would be far less likely to happen.

As it was, human thinking time had been largely eliminated for the trading system, and the algorithms were essentially trading with themselves; this already complex system had become even more tightly coupled. With the rise of computers, they could also access more buyers and sellers than before; the complex system was now more interrelated and, consequently, more intractable. Finally, computers can execute far more trades in a minute than a human could; the system has become far more dynamic.

Increased coupling, decreased tractability, and increased dynamic activity all came together on this day and created a positive feedback loop that could have tanked the market even further if the automatic circuit breaker system hadn't kicked in and created a precious five second gap that allowed the algorithms to reset their parameters and arrest the downward price spiral. It remains to be seen what could have happened if that automatic circuit breaker system hadn't worked.

As well as a massive increase in the complexity of the market, we are again faced with the structural problems of the global financial system, albeit to a lesser degree. While the Flash Crash certainly had implications outside of the US market, its brevity and its locality to one particular section of the US markets may have been what limited the scope of the damage internationally. There was still enough of a SYS2, 3, and 4 to intervene when it became clear that a domain shift had happened from the Complex to the Chaotic domain. The Regulatory Body also responded in a way appropriate to the new domain: act-sense-respond. The act was the activation of the automatic circuit breaker creating a five second stop in trading that broke the cycling positive feedback loops between the algorithms.

As in so many financial markets, the Regulatory Bodies of these systems are not particularly powerful. Arguably, the crash might have been less likely to happen if US regulators had taken a firmer line on the illegal practices used by HFTs. As well as spoofing, other techniques relied on increasingly rapid communication links to spot a movement in one market and then act in a different market, knowing that that market's movement would correspond to the movement of the first market. If you can get the order into the second market a millisecond before it receives the news that the first market has moved, you can make a profit. As soon as this began to happen, trading turned into technological warfare; it didn't matter what you knew, you just needed to be able to react in the market quicker than anyone else.

Many of these types of practices were cracked down on by the regulators in the wake of the Flash Crash, but it remains to be seen if this will be enough to prevent these bizarre patterns of activity from recurring again in the future.

Chapter 4.14

Alphafold 2[1,2]

Proteins are remarkable molecules. We can think of them as the machinery that keeps our cells alive and, in so doing, keeps us alive. The purpose of our DNA is to act as a template for creating proteins. When our cells want to make proteins, the relevant section of DNA is transcribed into mRNA (messenger RNA), which is then translated by ribosomes into a protein. It is worth pointing out that the machines that do the transcription and translation are themselves made of proteins acting together.

We know that particular sequences of DNA encode particular protein building blocks, known as amino acids. There are 20 known amino acids, and these are strung together like beads on a chain to form the basic structure of a protein. Some proteins are composed of as few as 11 amino acids, and the longest protein is known to have 34,350 amino acids. The average number of amino acids in human proteins is approximately 500. As you can imagine, with 20 potential amino acids to use and a chain with space enough for 500 of these amino acids, the number of potential combinations is astronomical.

The ability of proteins to act as cellular machines, sometimes in quite a literal sense, is based not only on the components (the amino acids on the chain) but on the overall three-dimensional structure of this chain of components. Because of the different physical and chemical properties of different amino acids, the chain doesn't remain in a straight line. For instance, some amino acids have a positive charge, and some have a negative charge. Two amino acids on the chain with opposite charges might be pulled closer together, thus deforming the chain. This deformation may now bring other amino acids into proximity, and this will change the attractive forces between them. Electrochemical charge is not the only variable characteristic between amino acids, but it illustrates the point.

Because these charge-related deformations are based on reducing the protein's energy state, we can say that the way proteins deform naturally follows an energy gradient in keeping with an increase in local entropy. It just so happens that this reduction to a lower energy state causes the molecule to fold up.

The initial folding creates helices and sometimes flat sheets of amino acids from the initial straight chain of amino acids. This is the protein's secondary structure. These helices and sheets also exert an attractive or repulsive effect on other parts of the amino acid sequence, and so the protein folds even

Designing and Managing Complex Systems. https://doi.org/10.1016/B978-0-323-91609-7.00026-3

further. This is now the tertiary structure of the protein. Finally, different folded proteins subunits can come together to form a larger protein. Hemoglobin, the protein that carries oxygen in the blood, is made up of four separate protein subunits that come together to form one larger protein. This self-assembled structure made of subunits is the protein's quaternary structure.

It is not enough to know the sequence of amino acids in a protein when it comes to biology and medicine. We must know how the protein folds for that information to be of use. Some of these folded structures are immensely complex. Proteins embedded into the walls of our cells can act as receptors and actually change their shape when stimulated by another chemical. The shape changing may then affect the structure of other proteins and start a chain reaction of some sort. Other shape changes may open a gateway in the cell membrane that allows chemicals to flow inward or outward.

Just about every biochemical process that occurs within our bodies is based on the activity of proteins. In some diseases, proteins can actually act as infectious agents, with misfolded proteins affecting normal proteins and causing them to change their shape so that they can no longer perform their function. This is the mechanism of prion diseases such as Creutzfeldt–Jakob disease, prions being the name given to misfolded, infectious proteins.

The challenge for modern medicine has been to understand how a straight chain of amino acids can turn into a highly complex folded structure. By understanding the structure of proteins, we can better understand how they work and, more importantly, design drugs that will modify their activity to treat disease.

Since proteins were originally identified, various methods have been used by the community of structural biologists specializing in this field. Coaxing proteins into a crystal formation meant that they were amenable to examination by X-rays, which helped clarify their structures. Newer techniques involve embedding proteins into a frozen substrate and using electron microscopy to probe their structure. Over the years, this sort of research has helped expand the database of protein structures and has given scientists important insights into how these proteins function in healthy individuals and in those with an illness related to protein dysfunction, of which there are many. In fact, most genetic diseases only manifest themselves because the incorrect genetic code leads to an incorrect protein sequence, which then has the incorrect three-dimensional structure.

Research into protein folding has taken a huge leap forward in the last few years. This change potentially means altering the basis for how the system designed to uncover the complex structures of proteins will work in the future.

With the goal of being able to work out protein structure, structural biologists have built up an array of techniques in SYS4KNOWLEDGE. Some proteins are harder to crystalize than others. When a researcher develops a new technique that can help coax a protein into a crystal structure to be analyzed, information gets added to SYS4KNOWLEDGE. Up until recently, that has been how structural biology has progressively determined the shapes of proteins. The shapes are also added to SYS4KNOWLEDGE. It is estimated that our planet's ecosphere includes about 40 million different types of protein. We currently have structures for about 50,000 of them.

The 21st century has seen the rapid evolution of deep learning computer algorithms. By absorbing knowledge already collected, deep learning systems can start to make inferences about things that are not currently known. Each successive step then provides more knowledge that it can draw on in the future to make conclusions.

Since the late 1990s, an annual competition has happened between structural biologists. They are given the amino acid sequence of a protein and have to use whatever techniques are at their disposal to try and work out its structure. One particular deep learning system has started delivering remarkable results in the last few years. The latest incarnation of this system is Alphafold 2.

4.14.1 Analysis

Deep learning algorithms redesign the system structure so that instead of SYS4KNOWLEGDE simply becoming a repository for knowledge, it also becomes a source for generating knowledge. The way to imagine this is as a two-way connection between SYS4LEARN and SYS4KNOWLEDGE. Alphafold 2 uses the structure of previously studied proteins to make inferences about the folded structure that will be created from a sequence of amino acids. For example, by looking at different proteins that occur in certain species, the program can identify where certain changes in the amino acid sequence have occurred as that animal has evolved. If the new sequence is different by two amino acids, it may be that those two amino acids were close together in the original structure. Algorithms can also use knowledge from physical chemistry to assess the likelihood that a particular bond between atoms results in part of the chain folding in one particular direction.

By going through this process iteratively, SYS4KNOWLEDGE and SYS4LEARN can build fresh knowledge through past experience. In the current early stages of this process being used, the new knowledge will need to be validated, but as the database of known protein structures increases, the accuracy of predicted structures will likely increase too.

The interaction between human-driven learning and machine-driven learning will likely be a significant part of systems engineering going forward. Already, large data-rich organizations mine this data to gain competitive advantages, as we have seen in the Netflix case study. However, applying new learning methods to generate more knowledge could be a rare example of successful positive feedback loops occurring within a system. When SYS4LEARN is adapted to take advantage of the growth in

SYS4KNOWLEDGE, the implications are exciting. For the world of structural biology, this fundamental rethinking of how the learning and knowledge subsystems are used could propel the field forward very quickly and open up a wide range of therapeutic approaches which are currently impossible.

For the design and management of systems in general, if we can harness the technology efficiently, the growth of these deep learning algorithms is going to make the SYS4LEARN/SYS4KNOWLEDGE feedback loop a valuable resource for all manner of different systems.

Part V

Conclusion

Chapter 5.1

Consilience with the arts

Chapter outline

5.1.1 Introduction

Throughout this book, we have tried to identify where patterns happening in one type of system may give us clues to what is happening in another type of system. To this end, we've looked at types of social systems, physical systems, informational systems, and many others besides. There is one glaring gap in this array of academic disciplines: the humanities.

In his book, "Consilience: The Unity of Knowledge," Edward O. Wilson dedicates a chapter to emphasizing the different ways that the humanities crossover into other areas of academic inquiry.[1] C.P. Snow, in his way, decries the fact that the sciences and the humanities are increasingly being seen as two different and independent avenues of thought.[2] For myself, I would agree with Snow that this is happening, and I would agree with Wilson that we all benefit from recognizing the unity of knowledge and from fighting the urge to remain in the intellectual comfort zones created for us by the modern drive toward specialization.

However, these good intentions still leave me in a quandary. My education was almost entirely in the sciences, and while I think I have an appreciation of humanities, I recognize my limitations when it comes to talking about them. With this in mind, there are still a few areas where even my meager abilities can recognize that some phenomena in the humanities seem to be replicated in other academic spheres. I will approach this subject by looking at music and visual art specifically.

5.1.2 Simple patterns and complex sounds

The American minimalist composer Steve Reich was an early proponent of electronic music. Perhaps not the electronic music that we would consider today as being popular in dance clubs. Rather, he was building machines that

Designing and Managing Complex Systems. https://doi.org/10.1016/B978-0-323-91609-7.00027-5

could modify tones, create pulses, and interweave different strands of music to create something original.

In the 1960s, he began to work on phase music. It was a musical idea that had been explored to an extent previously, but it was Reich who popularized it through several pieces composed in light of its philosophy.

Phase music is created by taking two or more identical scores of music and then gradually changing the speed of one relative to the others. What started as a synchronized, uniform collection of notes gradually increases in auditory complexity as the lines of music move out of phase with each other and then back into phase again, such as with his work Piano Phase. The complexity of the underlying music affects the overall complexity as the lines move out of phase, but the effect can be achieved with looped samples of music in a similar manner.

The effect is quite startling. Even as the lines move out of phase, certain repeated sequences within the repeated lines seem to synchronize and create new, briefly lived regular pulses that soon blend into a counterrhythm as the phase shift increases. Just by adjusting the phase between two identical lines of music, surprising complexity can arise.

More widely within the minimalist movement, short repeating phrases are interposed with other short phrases to create a complex auditory experience from relatively simple snippets of music. The minimalist movement grew during the 20th century and became a major part of the contemporary classical music world. The conclusions of the minimalist composers about the power of simple phrases interwoven in a regular way foreshadowed the conclusions of contemporary scientists on how complexity can arise from the combination of simple rules.

Cellular automata are computer programs that display their output on a large grid, the size of which can be determined by the operator. The squares on the grid are either white or black. Rules can then be set which determine what will happen next. For example, a rule may say that if a square is white and four or more of the eight squares around it are also white, in the next step, it will become black. Conversely, if a square is black and four or more squares around it are also black, it will become white. The starting pattern and the nature and number of the rules are up to the operator.

For many years, cellular automata have created much interest in the worlds of mathematics and computer science. Even with a single rule, after several generations, patterns begin to appear on the grid as cells switch from black to white and vice versa. Some rules will cause the automata ecosystem to increase in diversity, and others might wipe it out. The physicist Stephen Wolfram, who has done a lot of work on cellular automata, recognizes that different classes of rules will generate different types of patterns.[3] These classes map nicely onto the Cynefin framework; clear, complicated, complex, and chaotic. Some rules create uniformity, while others create complete randomness. Interestingly, the complexity of the patterns can be increased by having two very simple rules that interact with each other.

Are there similarities in the dynamics of the systems created by the minimalist composers and the computer programmers? I would say that the likely answer is yes, given how the regular, ordered changing of how components are interacting seems to create a more complex auditory or visual "ecosystem."

Cellular automata provide complexity theorists with a lot of research material, and some who are familiar with the field may have wondered why they haven't featured more prominently in this book. Again, there was a certain amount of cherry-picking which had to be done, but I also think the significance of cellular automata has been overstated in the field of complexity. Some might argue that the way cellular automata create shapes and patterns could be considered a form of visual art even though they are primarily studied from a scientific perspective. That being said, they do have some fundamental limitations.

While they certainly provide a visually engaging way of exploring complexity, they are, by nature, attuned to the observer. As humans, we attach significance to lines that seem to form or to patterns that we might recognize from nature appearing in the ecosystem of the cellular automata. They have also been used to reinforce the idea of emergence insofar as this complexity seems to appear from the repeated iterations of simple rules. It is mainly for this reason that I tend to be cautious about cellular automata. The patterns and shapes that we see forming before us as the computer program runs engage us because of how our brains are wired. When we see patterns of lines or rhythmic flashing shapes, we associate that with created structure. What we are doing is anthropomorphizing the output of a computer program, i.e., thinking that the program knows that what it is creating by repeating these rules is somehow structured. Again, the observer is "affecting" that which is being observed. If the data from the program were presented in its raw, binary form, we would likely not see any interesting patterns at all. It is only through the translation of this raw data into a visual form that we can see patterns that make sense to us as humans. This is not to say that they have not been important in the field of complexity science, but rather that we need to take a step back and realize that the "ecosystem" we have designed so we can watch these automata is necessarily influenced by how we interpret them. It is not just about the rules that determine their behavior. Seeing the musical score of a piece of phase music on the page is a very different from the experience of hearing it. That comparison highlights that there are likely more similarities between the seemingly remote worlds of minimalist contemporary classical music and cellular automata.

5.1.3 Complexity in the visual arts

Jackson Pollack, sometimes known as "Jack the Dripper," was an abstract impressionist artist of the mid-20th century. Famous for his so-called "drip" paintings, he would create large artworks by dripping and flicking paints of different colors onto large pieces of canvas. He would progressively add to these layers, varying the broad patterns and color selection. The canvases could be very large, the largest measuring 20 feet in width and 8 feet in height.

Observers of his technique noted the persistence of his physical movement around the canvas, which normally lay on the floor while he created it.[4] He would move around the painting with a paint can and a stick in hand, pouring or flicking paint at regular intervals. This would be repeated over a period of time until he was satisfied with the results. He would often trim the canvas after he had finished painting it to focus on the area that he felt was most aesthetically pleasing.

Critics of the day were divided about his work, but the art market embraced his drip style of paintings, and Pollock was successful in his lifetime. Since his death, the secondary market for his works has seen them fetch even higher prices.

The appeal of Pollock's work has not always been easy to determine. The question of what it was in this seemingly chaotic mix of lines, dots, and shapes that made these huge paintings aesthetically pleasing to many people wasn't easy to answer. It wasn't until the latter part of the 20th century, almost 50 years after the artist had died, that science began to uncover some previously hidden characteristics of his work.

Many of Pollock's drip paintings share a curious feature. Whether viewing the entire painting from a distance or looking at a smaller part of it from much closer, the overall structure appeared to be the same. It is not exactly the same in terms of precise patterns and colors, but the general level of complexity of the interaction between the lines and patterns seemed to be the same. If shown a picture of the entire painting and a picture of a small part of it, often it would be difficult to tell which was which.

Analysis of these paintings shows that, like viable systems, they are recursive; the same patterns exist across different scales. More specifically, the paintings share the same properties as mathematical structures called fractals.

Fractals are mathematical functions that are highly recursive. If plotted on a graph, complex patterns repeat endlessly as you zoom in on the graph to a greater and greater extent. Curiously, fractals were not recognized in mathematics during Pollock's lifetime, having only been defined during the 1970s. One of the early mathematicians to recognize these phenomena was Benoit Mandelbrot, and the intricate, almost psychedelic, visual graphs that represent one group of these recursive, fractal mathematical structures are called the Mandelbrot Set. As you zoom closer and closer into the graph, the complexity doesn't decrease, and new recursive structures continue to reveal themselves at every level. People familiar with Pollock would say that he knew when a painting did or didn't have "something," so we can only speculate whether he was looking for this type of recursive structure even if he could not verbalize that this was the characteristic he was striving for.

Aside from mathematics, recursive fractal patterns occur widely in nature.[5] When we look at trees, from the scale of the large branches to the scale of the twigs, the branch patterns appear to follow a fractal pattern. The florets of broccoli also have a recursive fractal pattern. Fractal patterns occur in geological structures, the leaves of plants, the shells of invertebrates, and the branches of neurons within the brain.

Analysis of Pollock's paintings confirms them to be highly fractal, and it has been suggested that the aesthetically pleasing nature of his work could be due to our innate appreciation for fractals, given their widespread existence in the natural world. Such is the extent of the fractality, especially in his later work, assessing this parameter has become part of authenticity tests for paintings that may have been made by him but whose provenance is unconfirmed.[6]

If that feature of Pollock's paintings had been recognized earlier, perhaps it would have hastened work in the field of fractal geometry, given that the dynamics are essentially the same. Interestingly, it was an academic who was dual trained in both physics and art, Richard Taylor, who recognized this property in Pollock's work.[7]

Chapter 5.2

Conclusion

The love of complexity without reductionism makes art; the love of complexity with reductionism makes science.

Edward O. Wilson, biologist.

Warren McCulloch was having difficulty with a technical problem. It was 1947, and he and his colleague, Walter Pitts, were trying to develop an electronic device that would help blind people to be able to interpret text by using a scanning device that would turn letter shapes into sounds. The problem was how could such a device deal with text of different sizes. He went to another one of the founding fathers of cybernetics, Norbert Wiener, to get his advice.

Discussing the problem in the common room, they decided that if they could rig up a photoelectric cell that would be cross-linked with other cells and oscillators through which the text could be scanned, the tones produced would be the same no matter what size the text was. The technical schematic they had drawn together appeared to be a complex, interlinked set of lines showing the connections between various parts of the scanning equipment. After they had finished, they set their schematic aside and moved on to other matters.

Gerhardt von Bonin, a renowned neuroanatomist, entered the common room and saw McCulloch and Wiener's schematic. Picking it up, he asked, "who has been trying to draw the brain's visual cortex?" At that point, McCulloch and Wiener realized that the technical schematic for their sophisticated visual scanning device was essentially the same as the wiring within the brain that allows us to interpret visual stimuli. When they calculated the optimum scanning rate for their device, they found that it had the same frequency as human brainwaves when the brain is monitoring the environment.[1]

Throughout this book, we have repeatedly seen how dynamics in one field seem to be replicated in another. They might operate at different scales and with different components, but the underlying dynamics are incontrovertibly the same. It was the cybernetic movement of the 20th century that first started putting a shape to this phenomenon and began to recognize how we could take advantage of findings in one field to benefit efforts in another. We've built on this idea of consilient dynamics across fields and across scales as well. Stafford Beer was the first person to consider that the workings of the human

Designing and Managing Complex Systems. https://doi.org/10.1016/B978-0-323-91609-7.00028-7

nervous system could give us ideas for managing complex systems. By including contemporary findings from the field of neuroscience, we have seen how the original Viable System Model can be extended to take advantage of the possibility that the new dynamics we've uncovered in the nervous system can offer us insights about designing and managing organizations.

This idea of consilience clearly offers us opportunities but, by ignoring its importance, we might end up taking a longer route toward a particular solution to a problem than is necessary. As I have been involved in the developing field of Resilience Engineering, I sometimes feel that we are reinventing some of the material that has already been uncovered through cybernetics. Similarly, the drive toward improving industrial safety has uncovered findings that are just as relevant in the drive toward improving operational management techniques. The systems may be different but they are operating in a very similar way.

In the same way as we ignore the lessons of history at our peril, we waste time by failing to recognize the commonality that exists between systems that might be operating in very different arenas. If we open our eyes to this possibility, embracing the idea of consilience might help us avoid pointless replication of effort.

It is inevitable that a book that takes as broad a view of the evolution of systems over the timespan of the universe risks becoming too abstract. It is my hope that by grounding the principles we saw in Part 2 into the practical steps recommended in Part 3, I can then convince you of their value through analysis of the case studies in Part 4. Although the "technical" part of "sociotechnical systems" is becoming more pronounced in the 21st century as computers and similar technologies become more essential to their functioning, we risk opening ourselves up to another wave of "system accidents" (to use Charles Perrow's terms) if we ignore some of the basic dynamics that have been proven to be important. Conversely, we are now presented with a previously impossible opportunity by exploiting this new technology to make the SYS4KNOWLEDGE/SYS4LEARN connection within our systems as robust and generative as possible. The case studies on Netflix and Alphafold 2 are testaments to the possibilities of fully exploiting this connection. With each technological step forward, the opportunities for us to manage and exploit the sometimes ephemeral informational subsystems within our organizations grow proportionally. I believe that the increasingly complex systems of our 21st-century world will be built on this connection.

Like the management strategy best used with systems operating in the Complex domain, our approach to our general knowledge of systems needs to be iterative. Stafford Beer created a remarkably insightful model of systems based on his knowledge of human neurobiology from the 1960s. New findings mean that I can present a similar model but one that is updated by contemporary research. I have no doubt that future neurobiological research will present more opportunities to update this model.

What is useful to note is that as well as updated research in neurobiology, it has been possible to incorporate findings from the field of management studies, particularly the Cynefin framework. By seeing the similarities in how humans make decisions under uncertain conditions at the level of neurobiology and at the level of management, we have been able to identify points of connection that we can potentially exploit.

It would be impossible to state all the ways that the different phenomena we have covered in this book could be adapted to apply to the systems that you find yourself a part of. The system types are too variable and the possible dynamics too numerous even without considering that there are likely to be far more of these consilient phenomena than this book has looked at. I'm also painfully aware that my own lack of education in the humanities has left a hugely useful area of study relatively untapped. With this in mind, perhaps a reader with more experience in this area than myself will see opportunities and similarities that have eluded me.

One thing we haven't done as yet is to try and apply the models and concepts that we defined in Part 3 onto one especially important system. In Part 2, we looked at different classes of systems including the system of societies. In the context of that part of the book, we were doing this as a prelude to defining informational and sociotechnical systems more clearly. There is, however, another step we can take. We can start by asking what is the largest system that we are meaningfully a part of? I would argue that the largest system that we are a part of is that of our species. Although we are certainly part of the wider system of all life on earth, we don't really have as strong a connection with other types of organisms as we have with our own species. It is an interesting intellectual exercise to consider whether the system comprising all human beings on Earth constitutes a viable system? When we consider the case study of the Global Financial Crisis, the system elements that were missing in that situation are arguably the same elements that are missing when we consider our species as a whole. We are also faced with various disturbances to our system, e.g., climate change, pandemics, international conflicts, etc., that requires a global, coordinated response. Could the lack of any sort of Regulatory Body mean that it is inevitable that our response will fail or, even worse, that in reacting to a worsening situation, we exacerbate the problem as happened when the global financial system began to disintegrate in 2007? Exploring this particular perspective is outside the scope of this book, but given that we have seen on multiple occasions how these sorts of dynamics recur across scales and across different fields of human endeavor, it is likely that they will recur again in this wider system that comprises all the human beings on Earth. The question remains whether we have the necessary resilience to manage the system disturbances that we are experiencing, and those that are yet to come.

Time and again, however, we seem to return to the same mantra: structure, function, capability, adaptive capacity, and resilience. From the level of the neuron upward toward the level of an organization, and beyond into our societies, these phenomena recur again and again. Like the Mandelbrot set, repeating patterns exist across these wide scales and suggest that what works on one level may well work at another level. Similarly, the challenge for us as system designers and managers is to be able to take off the "specialization goggles" that modern educational methods have forced onto us to be able to see system dynamics, not in terms of physics, chemistry, biology, engineering, sociology, or the humanities. Instead, we can attempt to see systems as they exist in themselves without imposing our predetermined perspective on them. I frequently come back to the image of that first explosion of energy and matter that started our universe. Despite the increasing total entropy of our universe, we still get to exploit this tidal wave of matter and energy that was triggered at the birth of the universe as we surf down the face of it and use its energy and matter to create the complex systems that surround us.

As we look at all these classes of different systems, whether created by us or created by nature, we might be able to see that everything is just a dance of matter, energy, and information. If we can understand how this dance is coordinated in successful systems, no matter where in reality those systems exist, we can learn new ways of better designing and managing the systems that we create for ourselves.

References

Preface

1. Wilson EO. *Consilience*. New York: Alfred A Knoff; 2000.
2. Snow CP, Snow B. *The Two Cultures and the Scientific Revolution*. Vol. 960. Cambridge: Cambridge University Press; 1959.

1. Control and communication

1. Medina E. *Cybernetic Revolutionaries: Technology and Politics in Allende's Chile*. MIT Press; 2011.
2. Conway F, Siegelman J. *Dark Hero of the Information Age: In Search of Norbert Wiener, the Father of Cybernetics*. Basic Books; 2006.
3. Javier Livas Cantu. *Cybernetics, History & Origins 1994 [Video]*. YouTube; May 17, 2012. https://www.youtube.com/watch?v=XbcBWdeIcyY.
4. Wiener N. *Cybernetics or Control and Communication in the Animal and the Machine*. MIT press; 2019.
5. Wiener N. *Extrapolation, Interpolation, and Smoothing of Stationary Time Series: With Engineering Applications*. Vol. 8. Cambridge, MA: MIT press; 1964.
6. Abraham T. *Rebel Genius: Warren S. McCulloch's Transdisciplinary Life in Science*. MIT Press; 2016.
7. Husbands P, Holland O. The ratio club: a hub of British cybernetics. *Mech Mind Hist*. 2008:91–148.
8. Turing AM, Haugeland J. *Computing Machinery and Intelligence*. Cambridge, MA: MIT Press; 1950:29–56.
9. Ashby WR. *An Introduction to Cybernetics*. Chapman & Hall Ltd; 1961.
10. Ashby W. *Design for a Brain: The Origin of Adaptive Behaviour*. Springer Science & Business Media; 2013.
11. Time Magazine. *The Thinking Machine*. Time; January 24, 1949.
12. Pickering A. *The Cybernetic Brain*. University of Chicago Press; 2010.
13. Beer S. *Brain of the Firm*. 2nd ed. Chichester: John Wiley; 1981.

2.1 The simplification imperative

1. Koch K, McLean J, Segev R, et al. How much the eye tells the brain. *Curr Biol*. 2006;16(14):1428–1434.
2. Heinrich HW. *Industrial Accident Prevention. A Scientific Approach*. 2nd ed. 1941.
3. Reason J. *Managing the Risks of Organizational Accidents*. Farnham: Ashgate; 1997.

4. Lorenz E. *Predictability: Does the Flap of a Butterfly's Wings in Brazil Set off a Tornado in Texas?* Washington, DC: American Association for the Advancement of Science; 1972.

2.2 The language of systems

1. Woods DD. The theory of graceful extensibility: basic rules that govern adaptive systems. *Environ Syst Decisi.* 2018;38(4):433–457.
2. Eisenstadt P. How the buttonwood tree grew: the making of a New York stock exchange legend. *Prospects.* 1994;19:75–98.
3. Mukherjee S. *The Gene: An Intimate History.* New York: Scribner; 2017.
4. Ashby WR. *An Introduction to Cybernetics.* Chapman & Hall Ltd; 1961.
5. Boulding KE. General systems theory—the skeleton of science. *Manag Sci.* 1956;2(3):197–208.
6. Perrow C. *Normal Accidents: Living with High-Risk Technologies.* 1st ed. Princeton University Press; 1984.
7. Vandermuelen J. *Erik Hollnagel – to Err Is Human: The ETTO Principle [Audio Podcast]*; December 2008 (Interviewer). http://www.namahn.com/we-share/interviews/erik-hollnagel-err-human-etto-principle.
8. Snowden DJ, Boone ME. A leader's framework for decision making. *Harv Bus Rev.* 2007;85(11):68.

2.3 Classes of systems

1. Lester KV, Matson RA, Howell SB, et al. Speckle Observations of TESS Exoplanet Host Stars. II. Stellar Companions at 1–1000 au and Implications for Small Planet Detection. *Astron J.* 2021;162(2):75.
2. Bjorken JD, Paschos EA. Inelastic electron-proton and γ-proton scattering and the structure of the nucleon. *Phys Rev.* 1969;185(5):1975.
3. Johnson G. *Strange Beauty: Murray Gell-Mann and the Revolution in Twentieth-Century Physics.* London: Jonathan Cape; 2000.
4. Hoyle F. *On Nuclear Reactions Occurring in Very Hot Stars. Etc.* University of Chicago Press; 1954.
5. Kvenvolden K, Lawless J, Pering K, et al. Evidence for extraterrestrial amino-acids and hydrocarbons in the Murchison meteorite. *Nature.* 1970;228(5275):923–926.
6. Miller SL, Urey HC. Organic compound synthesis on the primitive earth. *Science.* 1959;130(3370):245–251.
7. Kauffman SA. *A World beyond Physics: The Emergence and Evolution of Life.* Oxford University Press; 2019.
8. Damer B, Deamer D. The hot spring hypothesis for an origin of life. *Astrobiology.* 2020;20(4):429–452.
9. Mushegian A. The minimal genome concept. *Curr Opin Genet Dev.* 1999;9(6):709–714.
10. Glass JI, Assad-Garcia N, Alperovich N, et al. Essential genes of a minimal bacterium. *Proc Natl Acad Sci Unit States Am.* 2006;103(2):425–430.
11. Stewart PA. The organization of movement in slime mold plasmodia. *Primit Mot Syst Cell Biol.* 1964:69–78.

12. Tero A, Takagi S, Saigusa T, et al. Rules for biologically inspired adaptive network design. *Science*. 2010;327(5964):439–442.

2.4 Neurobiological systems

1. Moriarty D. *Practical human factors for pilots*. Academic Press; 2014.
2. Hebb DO. *The Organisation of Behaviour: A Neuropsychological Theory*. New York: Science Editions; 1949.
3. Anderson JR, Bothell D, Byrne MD, Douglass S, Lebiere C, Qin Y. An integrated theory of the mind. *Psychol Rev*. 2004;111(4):1036.
4. Salvucci DD, Taatgen NA. Threaded cognition: an integrated theory of concurrent multitasking. *Psychol Rev*. 2008;115(1):101.
5. Kahneman D. *Thinking, Fast and Slow*. Macmillan; 2011.
6. Kahneman D, Slovic SP, Slovic P, Tversky A, eds. *Judgment under Uncertainty: Heuristics and Biases*. Cambridge university press; 1982.
7. Kahneman D. Maps of bounded rationality: psychology for behavioral economics. *Am Econ Rev*. 2003;93(5):1449–1475.
8. Gilovich T, Griffin D, Kahneman D, eds. *Heuristics and Biases: The Psychology of Intuitive Judgment*. Cambridge university press; 2002.
9. Lichtenstein S, Fischhoff B, Phillips LD. Calibration of probabilities: the state of the art. *Decis Mak Chan Hum Aff*. 1977:275–324.
10. Meehl PE. *Clinical versus Statistical Prediction*. Minneapolis: University of Minnesota Press; 1954.
11. Bishop MA, Trout JD. 50 years of successful predictive modeling should be enough: lessons for philosophy of science. *Philos Sci*. 2002;69(S3):S197–S208.
12. Grove WM, Zald DH, Lebow BS, Snitz BE, Nelson C. Clinical versus mechanical prediction: a meta-analysis. *Psychol Assess*. 2000;12(1):19.
13. Rasmussen J. Skills, rules, and knowledge; signals, signs, and symbols, and other distinctions in human performance models. *IEEE Transact Syst, Man, Cybernet*. 1983;(3):257–266.
14. Reason J. *Human Error*. Cambridge university press; 1990.
15. Anderson JR, Matessa M, Lebiere C. ACT-R: a theory of higher level cognition and its relation to visual attention. *Hum Comput Interact*. 1997;12(4):439–462.
16. Sporns O. *Discovering the Human Connectome*. MIT press; 2012.
17. Sandberg A, Bostrom N. *Whole Brain Emulation:* A Roadmap. *Technical Report #2008-3*. Future of Humanity Institute, Oxford University; 2008. www.fhi.ox.ac.uk/reports/2008-3.pdf.
18. Stickgold R. Sleep-dependent memory consolidation. *Nature*. 2005;437(7063):1272–1278.
19. McEwen BS, Seeman T. Protective and damaging effects of mediators of stress: elaborating and testing the concepts of allostasis and allostatic load. *Ann N Y Acad Sci*. 1999;896(1):30–47.
20. Juster RP, Bizik G, Picard M, et al. A transdisciplinary perspective of chronic stress in relation to psychopathology throughout life span development. *Dev Psychopathol*. 2011;23(3):725–776.

2.5 Sociotechnical systems

1. Trist EL. *The Evolution of Socio-Technical Systems*. Vol. 2. Toronto: Ontario Quality of Working Life Centre; 1981.

2. Civil Aviation Authority. *CAP 719: Fundamental Human Factors Concepts*. West Sussex, UK: CAA; 2002.
3. Taylor FW. *The Principles of Scientific Management*. Harper & brothers; 1919.
4. Nelson D. Taylorism and the workers at bethlehem steel, 1898-1901. *Pa Mag Hist Biogr*. 1977;101(4):487−505.
5. Jessop B. *Fordism and Post-fordism: A Critical Reformulation*. Routledge; 2005:54−74.
6. Rasmussen J. Risk management in a dynamic society: a modelling problem. *Saf Sci*. 1997;27(2−3):183−213.
7. Cook R, Rasmussen J. Going solid": a model of system dynamics and consequences for patient safety. *BMJ Qual Saf*. 2005;14(2):130−134.
8. Hollnagel E. *The ETTO Principle: Efficiency-Thoroughness Trade-Off: Why Things that Go Right Sometimes Go Wrong*. CRC press; 2017.
9. Vandermuelen J. *Erik Hollnagel − to Err Is Human: The ETTO Principle [Audio Podcast]*; December 2008 (Interviewer). http://www.namahn.com/we-share/interviews/erik-hollnagel-err-human-etto-principle.
10. Johnston N. The paradox of rules: procedural drift in commercial aviation. In: *Proceedings of the Twelfth International Symposium on Aviation Psychology*. April 2003:14−17. April.
11. Vaughan D. *The Challenger Launch Decision: Risky Technology, Culture, and Deviance at NASA*. University of Chicago press; 1996.
12. Dekker S. *The Field Guide to Understanding Human Error*. Aldershot, UK: Ashgate Publishing; 2006.
13. Hollnagel E. Cognition as control: a pragmatic approach to the modelling of joint cognitive systems. *IEEE J Syst, Man Cybernet*. 2002.

2.6 Consilient dynamics across scales

1. Einstein A. *Investigations on the Theory of the Brownian Movement*. Courier Corporation; 1956.
2. Carus TL. *Lucretius on the Nature of Things*. G. Bell and sons; 1882.

3.3 Capability and adaptive capacity

1. Hollnagel E. *FRAM, the Functional Resonance Analysis Method: Modelling Complex Socio-Technical Systems*. Ashgate Publishing, Ltd; 2012.
2. Hollnagel E, Woods DD, Leveson N, eds. *Resilience Engineering: Concepts and Precepts*. Ashgate Publishing, Ltd; 2006.
3. Woods DD. The theory of graceful extensibility: basic rules that govern adaptive systems. *Environ Syst Decis*. 2018;38(4):433−457.

3.4 Engineering resilience

1. Hollnagel E, Woods DD, Leveson N, eds. *Resilience Engineering: Concepts and Precepts*. Ashgate Publishing, Ltd; 2006.
2. Hollnagel E. *Safety-I and Safety-II: The Past and Future of Safety Management*. CRC press; 2018.
3. O'Reilly. *Velocity NY 2013: Richard Cook, "Resilience in Complex Adaptive Systems" [Video]*. YouTube; October 15, 2013. https://www.youtube.com/watch?v=PGLYEDpNu60.
4. Kahneman D. *Thinking, Fast and Slow*. Macmillan; 2011.

5. Woods DD, Branlat M. Basic patterns in how adaptive systems fail. In: *Resilience Engineering in Practice*. CRC Press; 2017:127–143.
6. Hollnagel E. The four cornerstones of resilience engineering. In: *Resilience Engineering Perspectives*. Vol. 2. CRC Press; 2016:139–156.
7. Woods DD. Essentials of resilience, revisited. In: *Handbook on Resilience of Socio-Technical Systems*. 2019.
8. Yu DJ, Schoon ML, Hawes JK, et al. Toward general principles for resilience engineering. *Risk Anal*. 2020;40(8):1509–1537.
9. Duchek S, Raetze S, Scheuch I. The role of diversity in organizational resilience: a theoretical framework. *Bus Res*. 2020;13(2):387–423.
10. McCauley C. Group dynamics in Janis's theory of groupthink: backward and forward. *Organ Behav Hum Decis Process*. 1998;73(2–3):142–162.
11. Dekker S. *Just Culture: Balancing Safety and Accountability*. Ashgate Publishing, Ltd; 2012.

4.1 Challenger and columbia

1. Presidential Commission on the Space Shuttle Challenger Accident. *Report of the Presidential Commission on the Space Shuttle Accident*. Washington, DC: U.S. government printing office; 1986.
2. Columbia Accident Investigation Board. *Columbia Accident Investigation Board Report*. Washington, DC: U.S. government printing office; 2003.
3. Risen C. *Allan McDonald Dies at 83; Tried to Stop the Challenger Launch*. The New York Times; March 9, 2021.

4.2 Wal-mart, FEMA, and hurricane katrina

1. Horwitz S. Wal-Mart to the rescue: private enterprise's response to Hurricane Katrina. *Indepen Rev*. 2009;13(4):511–528.
2. Sobel RS, Leeson PT. *Flirting with Disaster: The Inherent Problems with FEMA*. Cato Institute; 2006.
3. Edwards C. *The Federal Emergency Management Agency: Floods, Failures, and Federalism*. Cato Institute Policy Analysis; 2014:764.
4. Worthen B. *How Wal-Mart Beat the Feds to New Orleans*. Vol. 1. CIO Magazine; 2005.

4.3 Lake peigneur

1. Perrow C. *Normal Accidents: Living with High-Risk Technologies*. 1st ed. Princeton University Press; 1984.
2. Mine Safety and Health Administration. *The Jefferson Island Mine Inundation*. U.S. Department of Labor; 1981.
3. Moriarty D, Jarvis S. Trade-Offs as symptoms of mismatches between sociotechnical systems: a case study involving commercial aviation and air traffic control. In: 5th *Symposium on Resilience Engineering; Managing Trade-Offs*. February 2014:221.

4.4 The water temples of bali

1. Lansing JS, Kremer JN. Emergent properties of Balinese water temple networks: coadaptation on a rugged fitness landscape. *Am Anthropol*. 1993;95(1):97–114.

2. Suradisastra K, Sejati WK, Supriatna Y, Hidayat D. Institutional description of the Balinese subak. *J Litbang Pertanian*. 2002;21(no. 1):11–18.
3. Landmann A. *Subak: Balinese Water Temples and Water Management (Stephen Lansing) [Video]*. YouTube; August 9, 2013. https://www.youtube.com/watch?v=lNKrR27Crqw.

4.5 The global financial crisis

1. Financial Crisis Inquiry Commission. *The Financial Crisis Inquiry Report: The Final Report of the National Commission on the Causes of the Financial and Economic Crisis in the United States Including Dissenting Views*. Cosimo, Inc; 2011.
2. McKibbin WJ, Stoeckel A. The global financial crisis: causes and consequences. *Asian Econ Pap*. 2010;9(1):54–86.
3. Ramskogler P. Tracing the origins of the financial crisis. *OECD J: Financ Mark Trends*. 2015;2014(2):47–61.

4.6 Continental airlines

1. Brenneman G. Right away and all at once. *Harv Bus Rev*. 1998;76(5):162–173.

4.7 Three Mile Island

1. *The American Experience: Meltdown at Three Mile Island*. PBS [TV Program]; 1999.
2. Perrow C. *Normal Accidents: Living with High Risk Technologies*. Princeton, NJ: Princeton University Press; 1999.
3. Moriarty D. *Practical human factors for pilots*. Academic Press; 2014.

4.8 Cybersyn and the trucking strike

1. Medina E. *Cybernetic Revolutionaries: Technology and Politics in Allende's Chile*. MIT Press; 2011.
2. Beer S. *Brain of the Firm*. 2nd ed. Chichester: John Wiley; 1981.
3. Pickering A. *The Cybernetic Brain*. University of Chicago Press; 2010.

4.9 Biological and informational viruses

1. Abdelrahman Z, Li M, Wang X. Comparative review of SARS-CoV-2, SARS-CoV, MERS-CoV, and influenza a respiratory viruses. *Front Immunol*. 2020:2309.
2. Larson HJ. The biggest pandemic risk? Viral misinformation. *Nature*. 2018;562(7726):309–310.
3. Birch J, Interviewer). *Lisa Bortolotti: Misinformation. [Audio Podcast]*. London School of Economics; January 26, 2021. https://www.lse.ac.uk/lse-player?id=7bc2bf71-f025-4165-9093-13e3e0d84c1a.
4. Cinelli M, Morales GDF, Galeazzi A, Quattrociocchi W, Starnini M. The echo chamber effect on social media. *Proc Natl Acad Sci Unit States Am*. 2021;118(9).
5. van der Linden S, Dixon G, Clarke C, Cook J. Inoculating against COVID-19 vaccine misinformation. *EClinicalMedicine*. 2021;33.

4.10 Netflix

1. McFadden C. *The Fascinating History of Netflix*. Interesting Engineering; July 4, 2020. https://interestingengineering.com/the-fascinating-history-of-netflix.
2. Allen G, Feils D, Disbrow H. The rise and fall of Netflix: what happened and where will it go from here? *J Int Acad Case Stud*. 2014;20(1):135.
3. Moore K. *Netflix Library by the Numbers 2020. What's on Netflix*; January 1, 2021. https://www.whats-on-netflix.com/news/netflix-library-by-the-numbers-2020/.
4. Whittock J. *Netflix Originals Exec Reveals How Streaming Giant Makes Commissioning Decisions. Screen*; April 25, 2018. https://www.screendaily.com/news/netflix-originals-exec-reveals-how-streaming-giant-makes-commissioning-decisions/5128552.article.
5. Chow PS. Ghost in the (Hollywood) machine: emergent applications of artificial intelligence in the film industry. *NECSUS_Euro J Media Stud*. 2020;9(1):193–214.

4.11 Fukushima

1. The National Diet of Japan. *The Official Report of the Fukushima Nuclear Accident Independent Investigation Commission*. The National Diet of Japan; 2012.
2. International Atomic Energy Agency. *The Fukushima Daiichi Accident*. Vienna, Austria: International Atomic Energy Agency, 2015.
3. Murata A, Karwowski W. On the root causes of the Fukushima daiichi disaster from the perspective of high complexity and tight coupling in large-scale systems. *Symmetry*. 2021;13(3):414.
4. Casto CA. *Station Blackout: Inside the Fukushima Nuclear Disaster and Recovery*. Radius Book Group+ ORM; 2018.
5. Synolakis C, Kânoğlu U. The Fukushima accident was preventable. *Phil Trans Math Phys Eng Sci*. 2015;373(2053):20140379.

4.12 The Mumbai dabbawalas

1. Thomke S. Mumbai's models of service excellence. *Harv Bus Rev*. 2012;90(11):121–126.
2. Patel N, Vedula N. Dabbawalas of Mumbai. *White Pap Kenan-Flagler Bus School*. 2006;1(1):5–7.

4.13 Flash crash

1. CFTC S, SEC U. Findings regarding the market events of may 6, 2010. *Rep Staffs oCFTC SEC Joint Adv Comm Emerg Regulat Iss*. 2010;104.
2. Kirilenko A, Kyle AS, Samadi M, Tuzun T. The flash crash: high-frequency trading in an electronic market. *J Finance*. 2017;72(3):967–998.
3. Akansu AN. The flash crash: a review. *J Cap Mark Stud*. 2017.
4. Vaughan L. *Flash Crash*. Leya; 2021.

4.14 – Alphafold 2

1. Thornton JM, Laskowski RA, Borkakoti N. AlphaFold heralds a data-driven revolution in biology and medicine. *Nat Med*. 2021;27(10):1666–1669.

2. Jumper J, Evans R, Pritzel A, et al. Highly accurate protein structure prediction with AlphaFold. *Nature*. 2021;596(7873):583–589.

5.1 Consilience with the arts

1. Wilson EO. *Consilience*. New York: Alfred A Knoff; 2000.
2. Snow CP, Snow B. *The Two Cultures and the Scientific Revolution*. Vol. 960. Cambridge: Cambridge University Press; 1959.
3. Wolfram S. Cellular automata as models of complexity. *Nature*. 1984;311(5985):419–424.
4. O'Connor FV. Hans Namuth's photographs of Jackson Pollock as art historical documentation. *Art J*. 1979;39(1):48–49.
5. Mandelbrot BB. Stochastic models for the Earth's relief, the shape and the fractal dimension of the coastlines, and the number-area rule for islands. *Proc Natl Acad Sci Unit States Am*. 1975;72(10):3825–3828.
6. Taylor RP, Guzman R, Martin TP, et al. Authenticating Pollock paintings using fractal geometry. *Pattern Recogn Lett*. 2007;28(6):695–702.
7. Taylor R. Fractal expressionism—where art meets science. In: *Art and Complexity*. JAI; 2003:117–144.

5.2 Conclusion

1. Wiener N. *Cybernetics or Control and Communication in the Animal and the Machine*. MIT press; 2019.

Glossary and abbreviations

Adaptive capacity The ability to adjust functioning to manage disturbances.

Allostasis The process of adjusting structure and function in the face of repeated or persistent disturbances.

Autocatalysis Self-creation.

Autonomic Self-controlled.

Baryon A type of subatomic particle.

Bias An erroneous, persistent error mechanism in human decision-making due to hardwired heuristics.

Capability A requirement for a component to be able to fulfill its function. To have capability, the component needs input, output, control, preconditions, resources, and time.

Chaotic domain Cynefin domain requiring act-sense-respond management.

Clear domain Cynefin domain requiring sense-categorize-respond management.

Closed system A system that does not routinely exchange matter, energy, or information with its environment.

Complex domain Cynefin domain requiring probe-sense-respond management.

Complicated domain Cynefin domain requiring sense-analyze-respond management.

Component The simplest part of a system that can have a single assigned function. Components may be humans, machines, or procedures (including software, rules, and laws).

Connectome The unique arrangement of neurons, and patterns of connections (synapses) that occur in individual brains.

Consilience The convergence of knowledge from different fields of study that may appear to be different.

Constraint A way of directing the flow of energy so that we can use it to do work.

Cybernetics The study of control and communication in systems.

Cynefin framework A sense-making framework that helps determine the appropriate management strategy for systems existing in particular domains within the framework.

Disturbance Something that affects the operation of a system and requires a response. The disturbance can be generated outside or inside the system.

Domain Part of the Cynefin framework. Systems may operate in a particular domain and may move into other domains depending on conditions. These domains are Clear, Complicated, Complex, and Chaotic.

Domain shift The movement of a system into a different domain of the Cynefin framework. This is usually caused by some sort of disturbance.

Dynamics The normal patterns of activity within an active system.

Entropy The extent to which energy in a system is ordered or disordered. Total entropy in the universe increases as energy distribution becomes increasingly disordered.

Epistemic bubble An environment where only knowledge that is consistent with other knowledge in that environment is accepted, even if it is not consistent with reality.

Equifinality Biological and social systems operate differently from mechanical ones because system endpoints can be achieved from various starting conditions, i.e., two systems exhibiting equifinality can reach the same endpoint from different starting conditions.

Feedback A pattern of activity where preceding events either decrease or increase the level of activity within a system.

Function What a component, subsystem, or system is assigned to do.

Gene A portion of an organism's genetic material that acts as the blueprint for a particular protein.

Heuristic Hardwired cognitive shortcuts that allow organisms to make rapid decisions in time-limited situations.

Homeostasis The physiological use of negative feedback to keep certain biological parameters within set limits.

Joint cognitive system A control system designed to account for the different capabilities of human and automated controllers.

Knowledge-based performance The cognitive process of having to come up with a response to a completely novel situation by interrogating knowledge stores to try and find a response that could possibly work.

Law of requisite variety The more we expect our systems to do and the more resilient we expect them to be in the face of disturbances exerted by the dynamic environment in which they function, the more complex the system has to become.

Management cybernetics The application of cybernetic principles to the management of organizational systems.

Metasystem A class of subsystems required to coordinate activity between other subsystems.

Multiple goal-seeking A pattern of behavior where different systems, subsystems, or components are trying to achieve their own goals with insufficient coordination between them.

Natural sciences Contains all the branches of science aimed at describing phenomena seen in nature. This includes physics, chemistry, and biology.

Nucleosynthesis The creation of atomic nuclei either during the Big Bang or within active stars or other highly dense cosmological bodies.

Observer effect The act of observing something can change it. In this book, the observer effect also refers to the incomplete knowledge that an observer may have of the phenomenon they are observing and how the observer's knowledge state affects their interpretation.

Open system A system that routinely exchanges matter, energy, or information with its environment.

Operating body All of the SYS1 entities of a system.

Operating point A feature of the Dynamic Safety Model. It is the current operational state of an active sociotechnical system with respect to its surrounding failure boundaries.

Organic chemistry Chemical processes that involve molecules containing carbon atoms. Organic molecules are required for carbon-based life to exist.

Production rule The cognitive basis for decision-making. Formed by interconnected neurons that allow for pattern matching between stimulus and response.

Quark-gluon complex The indivisible system of the building blocks within subatomic particles and the force carriers that keep them constrained.

R-number The number of people that someone with a transmissible, normally viral, disease is likely to infect.

Recursion Similar patterns of arrangement occurring at different scales.

Regulatory body SYS3-5 of a viable system that enables the Operating Body to fulfill its functions.

Resilience The ability of a system to adjust before, during, and after expected or unexpected disturbances to maintain its function and take advantage of opportunities.

Rule-based performance The cognitive process that allows humans to respond to a stimulus by matching it to an appropriate production rule that dictates the best response.

Simplification This innate drive in humans attempts to better understand complex phenomena by reducing them to simpler parts. The risk is that in doing so, important features of the phenomena are lost.

Skill-based performance The cognitive process of being able to come up with an instantaneous response to a situation without having to engage in any conscious consideration. For example, maintaining the center of a lane while driving.

Sociotechnical system A system comprising humans, machines, and procedures.

Structure The general arrangement of components and subsystems.

Subsystem A collection of components within a system.

System A unified collection of components working toward a common purpose.

Systems thinking Considering the dynamics of entities across reality based on the arrangement and activity of their components and subsystems rather than in terms of what specific type of system it is.

Theoretic minima The minimum amount of energy and/or matter required for a system to be able to adequately fulfill its function. Normally only relevant to technical systems.

Tractability A measure of how understandable a system is. This can change in accordance with the observer effect depending on who is looking at the system.

Trade-offs The conscious choice to deviate from procedures because doing so is the only way that a function will be fulfilled. A common type is an efficiency/thoroughness trade-off.

Ultrastability The ability to maintain a steady operation despite any disturbances that might affect the system.

Abbreviations

ACT-R Adaptive Control of Thought – Rational
ANS Autonomic Nervous System
CDO Collateralized Debt Obligation
CIA Central Intelligence Agency
CNS Central Nervous System
ETTO Efficiency/thoroughness tradeoff
FEMA Federal Emergency Management Agency
FRAM Functional Resonance Analysis Method
HFT High-Frequency Trader
KB Knowledge-Based
MMR Measles, Mumps, and Rubella (vaccine)
MRI Magnetic Resonance Imaging
NASA National Aeronautics and Space Administration
PNS Peripheral Nervous System
RAS Reticular Activating System
RB Rule-Based
rVSM Revised Viable System Model
SB Skill-Based
SRB Solid Rocket Booster
VSM Viable System Model
WAD Work-As-Done
WAI Work-As-Imagined

Index

Note: 'Page numbers followed by "*f*" indicate figures "*t*" indicate tables and "*b*" indicate boxes.'

D

E